Environmental Chemistry

(2nd Edition)

By the same author:

- Chemistry of Natural Products: Amino Acids, Peptides, Proteins and Enzymes
- Polymer Science
- Engineering Chemistry
- Textbook of Organic Chemistry Vol.-I
- Textbook of Organic Chemistry Vol.-II
- Textbook of Organic Chemistry Vol.-III
- Oxidation in Organic Synthesis
- Medicinal Chemistry (2nd Edition)
- Introduction to Organic Spectroscopy
- Strategies for Green Organic Synthesis
- Green Chemistry (2nd Edition)
- Reduction in Organic Synthesis
- Environmental Science (2nd Edition)
- Manual of Environmental Pollutants Estimation in Air, Water and Soil

Ane's Chemistry Series:

- Alicylic Chemistry
- Antibiotics
- Organic Reaction Mechanism
- Steroids and Hormones
- Vitamins
- Terpenoids
- Drugs
- Alkaloids

Environmental Chemistry

(2nd Edition)

V.K. Ahluwalia

Former Professor of Chemistry
University of Delhi, India
Visiting Professor
Dr. B.R. Ambedkar Center for Biomedical Research,
University of Delhi, India

Ane Books Pvt. Ltd.

New Delhi ♦ Chennai

Environmental Chemistry
V. K. Ahluwalia

First Edition : 2008, Reprint 2008
Second Edition : 2013, Reprint 2015, 2016, 2017, 2018, **2020**

Published by

Ane Books Pvt. Ltd.

4821, Parwana Bhawan, 1st Floor, 24 Ansari Road, Darya Ganj,
New Delhi - 110 002, Tel.: +91(011) 23276843-44, Fax: +91(011) 23276863
e-mail: kapoor@anebooks.com, Website: www.anebooks.com

Branch Office:

- Avantika Niwas, 1st Floor, 19 Doraiswamy Road, T. Nagar,
  Chennai - 600 017, Tel.: +91(044) 28141554, 28141209
  e-mail: anebookschennai@gmail.com, rathinam@anebooks.com

ISBN : 978-93-8116-296-5

Preface to the Second Edition

The enthaisatic response of the teachers and students for the first edition is greatly appecriated. The second edition has been completely revised, up grated and enlarged. On the basis of suggestions received, Three new chapters on Noise pollution, Environmental pollution and health and Energy and Environment have been added. Besides these a set of twelve experiments which are required for the laboratory work by the students have been incorporated in part II of this book.

It is hoped that the second edition will be extremely helpful to all concerned.

The author takes the opportunity in expressing his thanks to Dr. Suman Dudeja for her help and suggestions. The help rendered by Mr. Sunil Saxena of Ane Books Pvt. Ltd. is for the publication of the second edition is greatly appreciated.

V.K. Ahluwalia

Preface to the First Edition

Environment has become a matter of concern for all. With this idea, Environmental education has been implemented at the various levels of teaching programmes. This includes, knowledge of environment at the school, college and postgraduate levels.

There has been a tremendous increase in the developmental activities all around the globe. All these activities have a great impact on the environment. This, of course, has been very alarming in the last few decades. There is thus a need for conservation of the environment.

The Bhopal gas tragedy in December 1984, the worst chemical disaster disaster to date-fill, has focused attention towards the importance of environmental education. Many environmental issues, have grown in size and complexity, and now threaten the very survival of mankind on earth.

The objective of environmental education is to create an awareness in the public including students at all levels, about the importance of the protection and conservation of our environment and the need to restrain or curtail human activities which lead to indiscriminate release of pollutants into the environment.

With a view to achieve the above objectives, the NCERT and the university grants commission have recommended to educate the students at all levels to the various aspects of environmental knowledge. Keeping this in mind, this book has been developed to make the users aware of the different aspects of the environment and the related issues. The measures to prevent the environment degradation have also been incorporated.

Inspite of the best efforts, there might be certain points left out to make it a perfection. The author will be greateful to welcome such suggestions from its readers, to make the next edition more focused.

V.K. Ahluwalia

Contents

ENVIRONMENTAL CHEMISTRY PART–I

EXPERIMENTS PART–II

Part–I

Environmental Chemistry

Environmental Chemistry

GENERAL INTRODUCTION

Environmental Chemistry is an interdisciplinary science involving all branches of science, i.e. chemistry, physics, life sciences, agriculture, etc. It is the science of chemical phenomenon that occurs in the environment and deals with the study of the various sources, reactions and the fate of chemical species in the water, soil and air. In fact, environmental chemistry is a major and most important aspect of environmental studies. It deals with the affects of human activities upon the segments of the environment such as atmosphere, hydrosphere, lithosphere and biosphere.

It is well known that man has made tremendous progress in various fields of science and technology. Chemistry is responsible for medical revolution in the nineteenth century, in which drugs and antibiotics have been used to cure diseases, including life threatening diseases. These advantages brought about by chemistry, have resulted in the average life expectancy rising from about 47 years in the year 1900 to about 80 years in the 2010. The world's food supply has increased manyfold because of the development of chemicals (e.g. agrochemicals like fertilizers, insecticides, fungicides, herbicides etc.) that increase the yield and protect the crops. Virtually, in all aspects of material life (e.g. transportation, communication, clothing, human health etc.) chemistry has played a vital role in improving the quality of lives of billions of individuals who inhibit the planet. All these developments have come at a price, which has resulted in the pollution of the environment. The chemical industries have been responsible for releasing hazardous chemicals (and waste products) into the environment. Chemistry is responsible for a number of environmental problems.

(*i*) Chlorofluorocarbons (CFCs). These are a group of organic compounds made up of carbon, chlorine and fluorine and are extensively used as refrigerants, as solvents, in electronic industry and as propellants in aerosols. These CFCs when released in the atmosphere are responsible for the destruction of ozone layer in the stratospheric layer. The ozone layer absorbs the harmful ultraviolet radiations from the sun. In the absence of ozone, these harmful radiations reach the earth and may cause skin cancer and other ailments.

***(ii)* Pesticides.** These are chemical substances which are used to kill unwanted insects, fungus, rodents, etc., and are helpful to increase production of food. However, these insecticides pose a health risk and damage the environment. For example, widespread use of DDT has resulted in pollution of croplands and a large number of pests have become resistant to it. Also, DDT enters the food chain and gets accumulated in the fatty tissue of animals and is responsible for a number of problems.

***(iii)* Dioxins.** During the Vietnam war (1970's), the US Military sprayed about 42 million litres of the herbicide, 'agent orange' [a mixture of 2,4-dichlorophenoxyacetic acid (2,4-D) and 2,4,5-trichlorophenoxyacetic acid (2,4,5-D)], a defoliant to destroy the forest cover, so that the communist forces do not find place to hide themselves. As the herbicide 'agent orange' was contaminated with dioxins, it created havoc in the life of millions of victims of the war and other residents of the area. Dioxins are the most important carcinogenic tested so far. The dioxins got absorbed in the skin of the population. The children born to such women were deformed with low IQ or mentally retarded. A number of them suffered from skin tumors.

***(iv)* Thalidomide.** In 1961, there was a scare in Europe about a drug, thalidomide, which was used by pregnant women to redress the effect of nausea and vomiting during pregnancy (morning sickness). The children born to women taking this drug, suffered from acute birth defects, in a number of cases in the form of missing or deformed limbs. About 10,000 such children were born worldwide.

***(v)* Methyl isocyanate (MIC).** A most tragic chemical accident in the history of mankind occurred in a pesticide manufacturing company in Bhopal, India on 2nd December, 1984. About 33 tons of MIC, an extremely poisonous gas used in the manufacturing of 'seven'—a pesticide leaked from a storage tank and spread like mist and cloud over the city of Bhopal. The accident believed to be caused by failure of various controls led to the death of over 25,000 people and severe injuries to more than 2,00,000 people.

***(vi)* Industrial discharges.** A number of industrial discharges have led to a number of environmental episodes some of these are:

(*a*) Love canal episode

(*b*) Minimata disease

(*c*) Itai-Itai disease

(*d*) Methaemoglobinemia

A discussion of these forms the subject matter of subsequent discussions. On the basis of what has been stated above it is clear that chemistry has played a major role in a number of environmental problems.

Chapter **1**

Environment

1.1 INTRODUCTION

The term 'environment' is viewed differently by different environmentalists. It means surroundings and is defined as the sum total of all living (abiotic) and non-living (biotic) components. The environment creates favourable conditions for the existence and growth of all living organisms including man. The environment has never been static or constant. It has been changing slowly, rapidly or drastically. These changes in environment has affected all living organisms including human beings. The changes in the environment may be beneficial or harmful. It is known that many species on earth (which could not adapt to the changing conditions of the environment) vanished or are on the brisk of vanishing. A typical example is the extinction of dinosaurs due to adverse environmental conditions. It is believed that the main cause of this was the appearance of tiny primitive mammals, which ate the eggs of dinosaurs. In this way, the environment was favourable for primitive mammals but unfavourable for dinosaurs.

Environment is comparatively a new movement in the world. In the developed and developing countries, environmental education is getting top national policies. The objective of environmental education is to make the public including the students at all levels aware of the importance of the protection and conservation of our environment and the need to restrain activities which lead to an indiscriminate release of pollutants into the environment. According to Haruhiko Tanaka of Hiroshima University, Japan (J. Indian Chem. Soc., 2002, 77, 537 – 538) environmental education not only aims at acquiring the content of knowledge related to environmental problems in the society, but also on fostering student's abilities and the attitude necessary for the precise judgement to solve the problems. This could be summarised by citing the so-called three aims (3A's) which should be adhered to if we want to improve our environment. There are awareness (environmental literacy), attitude (problem solving skills) and action (decision-making). It can be said that the ultimate aim of environmental education is to develop people's awareness, attitude, competence and values. This will enable them to be effectively involved in sustainable development at the local, national and international levels and guide them towards a more equitable and sustainable future. With this idea in mind, environmental education is now being taught at the school, college and university levels.

Human activities have been responsible for changes in the environment leading to its degredation. Some of the activities are:

- Deforestation due to cutting trees for wood and bringing forest land under cultivation. This is responsible for erosion of the soil. This activity has been going on for thousands of years.
- Killing gentle animals for food and the fierce animals due to safety reasons, has also been going on since ancient times and has led to the gradual extinction of animal species.
- The industrial and scientific revolution in the recent past has a tremendous effect on the environment. The various activities involved in this are mainly responsible for polluting the atmosphere by the discharge of a number of harmful and poisonous gases discharged from industrial establishment including power plants using coal. The activities have also resulted in polluting the water bodies with discharges from various industries.
- A large number of episodes have affected the environment. The most important is the London smog that killed about 4000 people in 1952, nuclear catastropes including dropping of nuclear bombs on Hiroshima and Nagasaki during the World War II that was responsible for the death of thousands of people and disabling millions of people for the rest of their lives. The leakage of methyl isocyanate gas (MIC) from the Union carbide factory in Bhopal, India, on 2 December, 1984. This was the world's worst ever industrial disaster.

1.2 COMPONENTS OF ENVIRONMENT

The environment is comprised of the following components:

(*i*) **Abiotic or Non-living components:** These include

Light	Altitude
Humidity and water	Latitude
Temperature	Seasonal changes
Atmospheric gases	Topography

(*ii*) **Biotic or living components:** These include plants (flora and fauna), Animals including humans, parasites and microorganisms, decomposers.

(*iii*) **Energy component:** It includes solar energy, geochemical energy, thermoelectric energy, nuclear energy, etc. Energy from other sources including radiation also plays an important role for sustaining life process.

1.3 TYPES OF ENVIRONMENT

The environment can be categorised as:

(*i*) **Natural Environment:** The components of natural environment are air, water, soil, solar radiation, land, forest, wildlife, flora and fauna, etc.

(*ii*) **Man-made Environment:** These include housing, agricultural implements, industries, dams, energy such as hydro, thermal and nuclear energy, etc.

1.4 SEGMENTS OF THE ENVIRONMENT

The environment is comprised of the following segments:

(*i*) Atmosphere

(*ii*) Hydrosphere

(*iii*) Lithosphere

(*iv*) Biosphere

Atmosphere. The cover of air that envelopes the earth is known as the atmosphere.

Hydrosphere. It is all of the water on or over the earth surface. It covers about 75 per cent of the earth's surface, either as oceans (salt water) or as fresh water and includes oceans, seas, rivers, lakes and streams, etc.

Lithosphere. The layer of rock that forms the outer part of the earth is called the lithosphere. It mainly includes soil, earth, rocks and mountains, etc.

Biosphere. The part of the earth's surface and atmosphere in which plants and animals can live. It comprises of lithosphere, hydrosphere and atomosphere.

Various segments of the environment form the subject matter of separate chapters.

1.5 ENVIRONMENTAL POLLUTION

Pollution of the three segments of the environment, *i.e.* the air (atmosphere), land (lithosphere) and water (hydrosphere) is referred to as environmental pollution. Some basic causes of environmental pollution are:

- Rapid industrial and technological development all over the world has resulted in the release of poisonous matter (gases and other by-products) into the air, water and land.
- The manufacturing units engaged in the production of drugs, antibiotics, dyes and agrochemicals (for example, pesticides, fungicides, herbicides, etc.) has added to the environmental pollution.
- Excessive use of fertilizers and pesticides for agriculture.
- Development of atomic and nuclear energy has resulted in the release of highly toxic radioactive materials into the environment.

- A large number of industrial effluents and emissions, especially toxic gases and other byproducts from cement, sugar, distillery, drug, paper and pulp, thermal power plants, nuclear power plants, metallurgical process, etc., are released into the environment.

Due to the above reasons and several other factors, the environment has become polluted to a great extent. The level of many pollutants has risen beyond the permissible limits and the environment is slowly becoming lethal to all organisms including man.

India, today is the seventh amongst the industrialised developing countries of the world. India, since independence has established a broad base for the manufacture of drugs, antibiotics, dyes, agrochemicals, plastics, polymers etc. It has made remarkable progress in the field of petroleum, atomic and nuclear energy. However, along with the progress our country has made, we have also contributed greatly towards environmental pollution.

It was only towards the end of the nineteenth century that people all over the world understood the concept of environmental pollution and the harm it can cause. In view of this, various governments (especially in the developed and advanced countries like USA, UK, France, etc.) have formulated laws to check the pollution of the environment. As a general rule all industries are required to treat their effluents and harmful byproducts they produce. These pollutants cannot be released into the environment. The Government of India has established the Ministry of Environment. Its basic objective is to formulate laws with a view to prevent environmental pollution.

All the causes of environmental pollution and ways to check pollution will be the subject matter of subsequent chapters.

1.6 COMPOSITION OF ATMOSPHERE

The atmosphere, as we know, is the region of air present above the earth's surface. It extends up to a height of about 500 km from the earth's surface. The atmosphere is capable of supporting life on earth and save it from the ill effects of the ultraviolet rays of the sun. The atmosphere consists of a number of gases like nitrogen, oxygen, carbon dioxide and some inert gases like argon. The actual composition of air near the sea level is given in Table 1.1 as major, minor and trace components. The composition is expressed in per cent by volume or ppm. The values in ppm is obtained by multiplying the per cent volume by 10^4.

Table 1.1 The Composition at Dry Air Near Sea Level

Components	***Amount***	
	% (by volume)	***ppm***
1. **Major Components**		
Nitrogen (N_2)	78.09	7,80,900
Oxygen (O_2)	20.98	2,09,400
Water vapour (H_2O)	~ 0.15	1,000 – 50,000
2. **Minor Components**		
Argon (Ar)	0.934	9,340
Carbon dioxide (CO_2)	0.032	320
3. **Trace Components**		
Neon (Ne)	0.00182	18.2
Helium (He)	0.000524	5.24
Methane (CH_4)	0.00018	1.8
Krypton (Kr)	0.00018	1.8
Nitrous oxide (N_2O)	0.000025	0.25
Hydrogen (H_2)	0.00005	0.5
Xenon (Xe)	0.0000087	0.087
Sulphur dioxide (SO_2)	0.0000002	0.002
Nitrogen dioxide (NO_2)	0.0000001	0.001
Ammonia (NH_3)	0.000001	0.01
Carbon monoxide (CO)	0.000012	0.12
Ozone (O_3)	0.000002	0.02
Iodine (I_2)	traces	traces

The approximate mass of the atmosphere is 4.5 to 5 × 10^{15} metric tons (about one millionth of the total weight of the earth). The temperature, pressure and density of the atmosphere vary considerably with altitude. Table 1.2 given various parameters.

Table 1.2 Various Parameters of Atmosphere

Parameters	***Amounts***
Total mass	4 – 5 × 10^{15} metric tons
Temperature[1]	–100°C to 1200°C depending on the altitude.
Atmospheric pressure	One atmosphere at sea level;
	3 × 10^{-7} atmosphere at 100 km above sea level.
Atmospheric density	0.0013 g/m³ at the surface of the earth. It
	decreases sharply with increasing altitude.

1. See Table 1.3.

1.6.1 Major Regions of Atmosphere (The Vertical Temperatures)

The atmosphere extends up to about 500 km above the earth's surface. The temperature varies as we go vertically from one region of the atmosphere to another. The major regions of the atmosphere, temperature of various regions and the main constituents of the region is given in Table 1.3.

Table 1.3 Major Regions of the Atmosphere

Region	*Height Range (km)*	*Temperature Range (°C)*	*Main Constituents*
Troposphere	0 – 11	15 to –56	O_2, N_2, CO_2, H_2O
Stratosphere	11 – 50	–56 to –2	O_3
Mesosphere	50 – 90	–2 to –92	O_2^+, NO^+
Thermosphere	90 – 500	–92 to 1200	O_2^+, O^+, NO^+

Besides the above major regions, the region of atmosphere lying between altitudes of 50 km to 100 km is called **ionosphere**. This region has appreciable levels of positive ions (e.g. O_2^+, O^+, NO_2^+, etc.) and electrons. These ions are formed due to the UV radiation from the sun. During night when there is no solar radiation, the ions combine with free electrons and give neutral species from which they originated. The division of atmosphere into various regions is shown in Fig. 1.1.

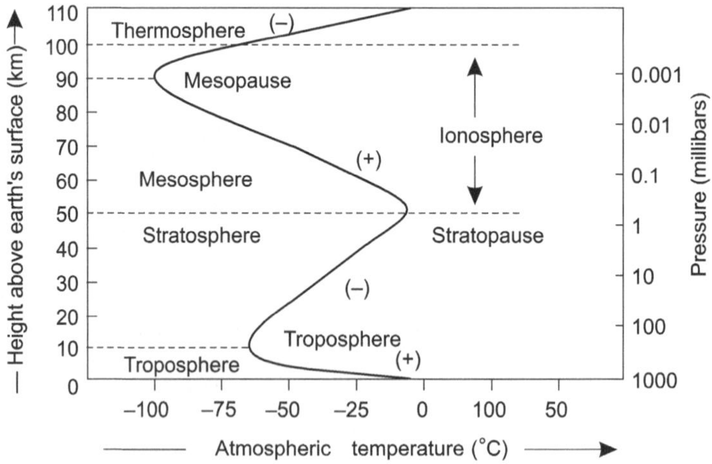

Fig. 1.1 Various segments of the atmosphere

(i) **Troposphere.** The region nearest to the earth, extending up to a height of about 11 km is called troposphere. About 70 per cent of the total mass of air is contained in the **troposhere**. In this region, the temperature steadily decreases from ground temperature of about 15° to about –56° (see Fig. 1.1). The fall in

temperature is due to the fall in pressure and the density of air. The change in temperature is known as lapse rate. The decrease of temperature with increasing height in troposphere is called **positive lapse rate**. This region is the most important zone of atmosphere for sustaining life on earth as it contains mainly oxygen and nitrogen. It also contains argon, carbon dioxide and traces of helium, neon, methane, hydrogen, nitrous oxide, carbon monoxide, and nitrogen dioxide, sulphur dioxide, etc. Troposphere is a dusty zone containing water vapours. This is the region of strong air currents and cloud formation and is also referred to as a **turbulent zone**. The composition of air in troposphere remains more or less constant provided there is no significant air pollution.

***(ii)* Stratosphere.** The region above the troposphere extending up to an altitude of about 50 km is the stratosphere. In this region, the temperature rises from –56° to –2°C. This increase in temperature is due to absorption of solar radiation. This region is rich in ozone, which absorbs the harmful ultraviolet radiations and prevents it from reaching the earth, and raises the temperture causing **negative lapse rate**. Besides ozone, this region also contains oxygen and nitrogen along with small amounts of water vapour. In stratosphere, ozone is formed from oxygen by a photochemical reaction in which energy from the sun decomposes the oxygen molecules into reactive oxygen, which in turn combines with molecular oxygen to form ozone.

$$O_2 + h\nu \rightleftharpoons 2O$$

$$O_2 + O \rightleftharpoons O_3$$

The content of ozone in the stratosphere remains more or less constant. This statement holds good only if pollutants like CFC's are not discharged into the atmosphere.

***(iii)* Mesosphere.** Next to stratosphere is the mesosphere, which extends upto an altitude of 90 km. It is characterised by low temperature (–2°C to –92°C) and very low atmospheric pressure. The region having temperature of about –92°C is the coldest region of the atmosphere. The decrease in temperature in this region is due to low levels of the ozone present, which is known to absorb UV radiations from the sun. The main constituents of mesosphere are the positive ions like O_2^+, NO^+, etc.

***(iv)* Thermosphere.** This region is above mesosphere and extends up to a height of about 500 km. The temperature in this region rises from –92° to 1200°C. The atmospheric gases (Oxygen and Nitrogen) in this region absorb the solar radiations in the far ultraviolet region and undergo ionisation.

***(v)* Ionosphere.** The region of atmosphere lying between the altitudes of 50 km to 100 km is the ionosphere. In this region, there are high levels of ions like O_2^+, O^+, NO^+ and electrons. These species are formed due to solar radiations. During the night, the UV radiations are not available and so the ions recombine with free electrons to give neutral species from which they originated.

***(vi)* Exosphere.** It is the uppermost region of the atmosphere and extends up to a height of about 1600 km. This region contains hydrogen and helium. It has very high temperature due to solar radiations. This region is referred to as the outer space. The division of the atmosphere into various regions is shown in a simplified way in Fig. 1.2.

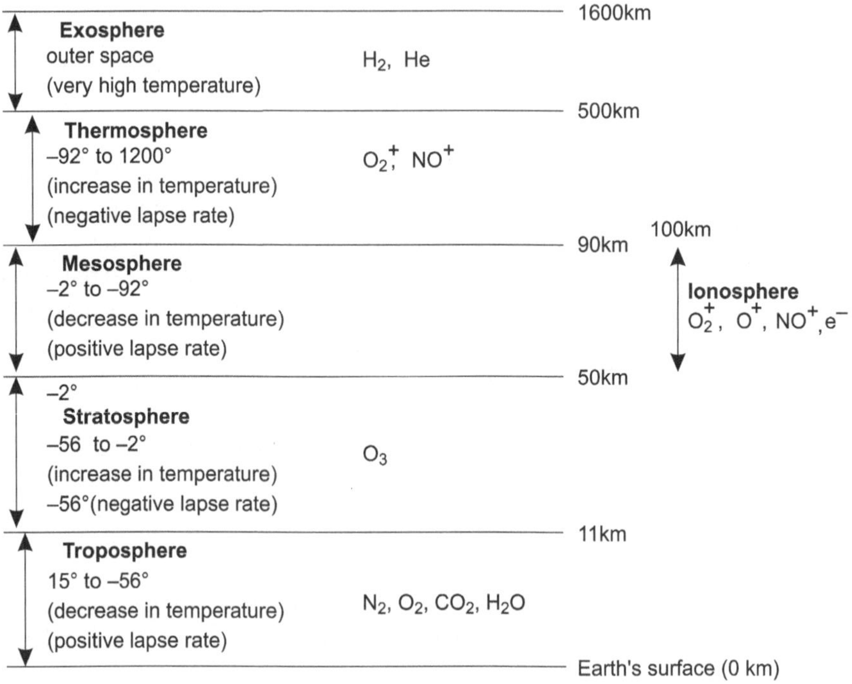

Fig. 1.2 Simplified representation of various regions of the atmosphere

1.7 HEAT BUDGET OF THE EARTH'S ATMOSPHERIC SYSTEM

The earth receives all its heat energy by radiation from the sun. This heat energy received by the earth is subsequently lost by reflection and radiation. The overall result is that the earth loses as much heat as it receives. Thus, on an average a near-steady state of average global temperature is maintained. Over the years, the earth has neither progressively cooled nor warmed and the temperature of the earth has remained more or less constant, although there have been some imbalances in local regions.

A decisive role in maintaining the radiation balance of the earth is played by the atmosphere. It is through the atmosphere that the solar radiations reach the surface of the earth and again it is the atmosphere through which the earth radiates or reflects back the solar radiations. On a rough estimate, it is believed that about 33 per cent of the solar radiations received from the sun are reflected back by the combined effect of the clouds, snow and ice sheets and the earth's surface. The reflecting capacity of the earth is called **albedo**. Of the remaining 68 per cent of the solar energy, the atmosphere absorbs about 16-17 per cent and 48-51 per cent is absorbed by the earth's surface. Ultimately, both the atmosphere and the earth's surface radiate back the absorbed energy as heat. It is the energy transport which plays a significant role in the earth's heat budget. The transport of energy takes place by the radiation of energy from the earth, conduction of energy through the interaction of atoms and molecules with the atmosphere and convection of energy through air circulation. The last two mechanisms, vig., conduction of energy through the interaction of atoms or molecules and convention to energy due to massive air circulation are responsible for loss of heat from the earths surface through transport to clouds and subsequent radiation from the clouds. About 67 per cent energy is radiated into the outer space. In this way, the heat gained and heat lost is balanced over the global scale. The temperature of the earth, on an average, is maintained around 15°C due to reabsorption of most of the outgoing radiation by carbon dioxide and water vapours. This combined effect of water vapours and carbon dioxide leads to the so called **greenhouse effect** (see Section 4.9.2).

The heat balance varies from one region of the earth to another; this aspect has a significant effect. The imbalance between the heat absorbed and radiated causes atmospheric and oceanic circulations to occur. This factor controls the climate and weather patterns of the globe.

The earth's heat balance is considerably affected by numerous human activities. It is well known that many industrial process are responsible for the discharge of large amounts of greenhouse gases like carbon dioxide, ozone, nitrous oxide, methane, water vapours and particulate matter into the atmosphere.

Deforestation, soil erosion and volcanic eruptions also have added to the misery. Since the carbon dioxide content of the atmosphere is increasing day by day at an alarming rate due to various human activities, the global temperature is also rising gradually. According to United Nations, Environmental Protection Agency (UNDP) a rise of 2°C temperature may disrupt the heat budget and will cause catastrophic consequences. It is estimated that by the year 2050, the temperature of the globe would rise by 1.5 to 4.5°C. The discharge of pollutants particularly greenhouse gases, if not controlled will be disastrous for the environment.

1.8 BIOGEOCHEMICAL CYCLES

For sustaining life in the biosphere, a number of nutrients are needed. The nutrients include the Big Six. These elements form the fundamental building blocks of life. These are carbon, nitrogen, oxygen, phosphorus, sulphur and calcium. Each element plays a special role in the organisms. These nutrients must be recycled so that once used they are made available again to the living organism. During cycling, interchange of nutrients occur among the biotic community and abiotic environment. The passing of nutrients between environment and organisms is referred to as **biogeochemical cycle** or **nutrients cycle** (bio for living and geo for atmosphere). Each nutrient is normally concentrated in a source, e.g. nitrogen is basically concentrated in the atmosphere. Such a source is called a **reservoir** or a pool. The biogeochemical cycles are of following two types:

(*i*) Gaseous Cycle. In such cases, the reservoir is the atmosphere or hydrosphere. Examples of this type include water cycle, carbon cycle, nitrogen cycle, oxygen cycle, etc.

(*ii*) Sedimentary Cycle. In these cases, the reservoir is the earth's crust. Examples of this type include phosphorus, sodium, potassium, magnesium and iron cycles.

Following discussion is devoted to some of the biogeochemical cycles like the carbon cycles, nitrogen cycle, oxygen cycle, phosphorus cycle and sulphur cycle. Water or hydrological cycle forms the subject matter of Chapter 2.

1.8.1 Carbon Cycle

Carbon is one of the most important constituent of all organic compounds. It is the building block of life. Carbon is an essential constituent of carbohydrates, proteins, fats, and nucleic acids. The main reservoirs of carbon are:

- Atmosphere, in which carbon is present as carbon dioxide.
- Oceans, in which carbon is present as carbon dioxide dissolved in water. In addition, it is also present as calcium carbonate in various seashells and in rocks at the bottom of sea.
- Land, on which it is present in some ores like dolomite and other carbonates.

The carbon cycle (or the so called carbon dioxide cycle) involves the circulation of carbon dioxide from the three sources as shown below in Fig. 1.3.

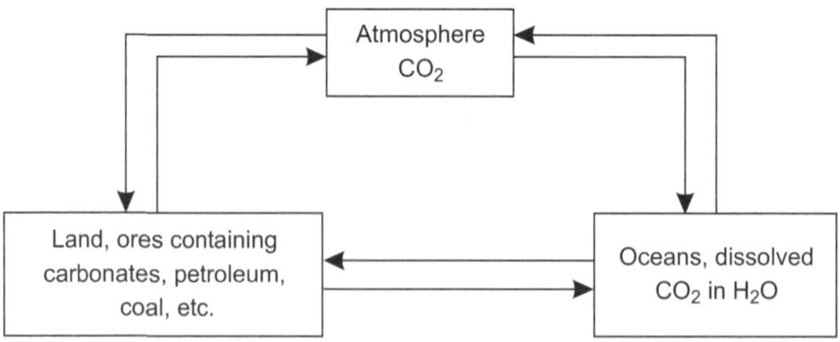

Fig. 1.3 Reserviors of carbon

The carbon or the carbon dioxide cycle involves the following process.

- Carbon dioxide from the atmosphere enters the plant through photosynthesis, where carbonhydrates are produced.

- From green plants, the carbon in the form of carbohydrates, etc., enter the animal and human bodies.
- The remains of dead animals and plants on putrification and bacterial decomposition release carbon dioxide back into the atmosphere.
- The atmospheric carbon dioxide gets dissolved in oceans, etc., by simple diffusion process. Marine algae and photosynthetic bacteria obtain carbon dioxide from water. Some of the dissolved carbon dioxide is also trapped to form lime stone (calcium carbonate) and other carbonate containing rocks.
- Carbon dioxide returns to the atmosphere by respiration, combustion of fossil fuels like coal, wood, petroleum, diesel, natural gas, gasoline etc.
- Weathering of carbonate containing rocks (particularly by water containing dissolved carbon dioxide as carbonic acid and by acid rain) releases carbon dioxide into the atmosphere. Carbon dioxide is also released into the atmosphere by volcanic eruptions.

The basic features of carbon cycle are depicted in Fig. 1.4 and Fig. 1.5.

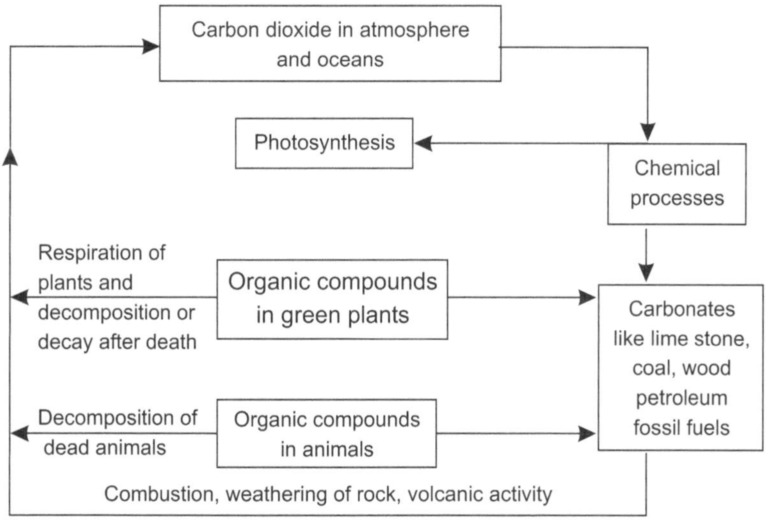

Fig. 1.4 Carbon cycle (flow sheet)

Note: Release of excessive amounts of carbon dioxide into the atmosphere is responsible for greenhouse effect and global warming (see Chapter 4).

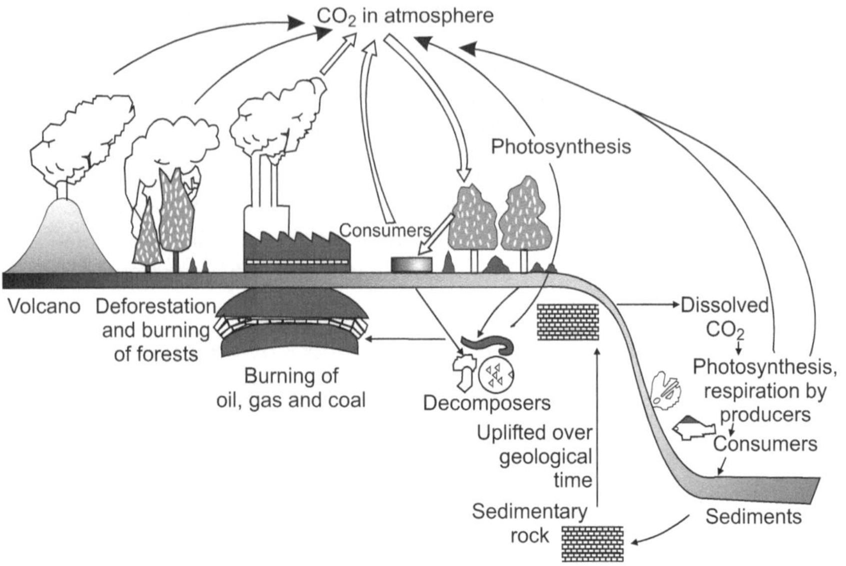

Fig. 1.5 Carbon cycles—various features

1.8.2 Nitrogen Cycle

In addition to carbon, nitrogen is an essential constituent of animals and plants, in which it is present as amino acids and proteins. Proteins are the building blocks of all living tissues. Though atmosphere is the main source of nitrogen in which it is present to the extent of 78–80 per cent. However, the elemental nitrogen cannot be used by most of the living organisms. Plants can take up nitrogen either as nitrates or ammonia in the form of salts. This is achieved by the fixation of atmospheric nitrogen either by a natural process or by a synthetic process.

In the natural process of fixation of atmospheric nitrogen, during thunderstorm and lightening, the gaseous nitrogen is converted into nitric oxide, which gets oxidised to nitrogen dioxide. Both the oxides of nitrogen are washed down as acid rain and reach the soil, where the formed nitric acid combines with salts in the soil to form the corresponding nitrates.

$$N_2 + O_2 \longrightarrow 2NO$$

$$2NO + O_2 \longrightarrow 2NO_2$$

$$4NO_2 + 2H_2O + O_2 \longrightarrow 4HNO_3$$

$$CaO + 2HNO_3 \longrightarrow Ca(NO_3)_2 + H_2O$$

The atmospheric nitrogen during thunderstorm may also combine with hydrogen in the atmosphere to produce ammonia, which is also washed down to the earth with rain and may combine with nitrates and sulphates in the soil to form ammonium salts.

$$N_2 + 3H_2 \longrightarrow 2NH_3$$

$$HNO_3 + NH_3 \longrightarrow NH_4NO_3$$

$$CaSO_4 + 2NH_4NO_3 \longrightarrow (NH_4)_2SO_4 + Ca(NO_3)_2$$

An alternative route for the natural process of nitrogen fixation is with the help of certain microorganisms, which can convert atmospheric nitrogen into ammonium ions. These organisms include nitrifying bacteria, (e.g. **aerobic azobacter** and anaerobic **clostridium** and symbiotic nitrifying bacteria living in association with leguminous plants and symbiotic bacteria living is non-leguminous root nodule plants (e.g. Rhizobium) as well as blue-green algae (e.g. Anabaena, Spirulina). Some plants are capable of directly taking up ammonium ions as a source of nitrogen. Alternatively, ammonium ions can be oxidised to nitrites or nitrates by specialised bacteria. For example, Nitroso-monas bacteria can promote oxidation of ammonia into nitrite, which is further oxidised into nitrate by the bacteria Nitrobacter. The nitrates, thus obtained in the soil are taken up by plants and converted into amino acids—the building blocks of proteins. The remains of plants decay and during the process, nitrogen is returned to the soil in the form of ammonia. Some of the nitrates (being soluble in water) find their way into the water bodies like rivers, oceans, etc. The denitrifying bacteria, e.g. Pseudomonas is present in soil as well as oceans; these convert the nitrates into elemental nitrogen, which escapes into the atmosphere, thus completing the nitrogen cycle.

In the synthetic process for the fixation of atmospheric nitrogen, nitrogen and hydrogen are made to react together in the ratio 1 : 3 at high temperature and pressure in presence of a catalyst. This forms the basis of Haber's process for the manufacture of ammonia. The ammonia thus produced is subsequently converted into fertilizers like ammonium sulphate, ammonium nitrate and urea. In another process, the nitrogen and oxygen of the atmosphere are made to combine at high temperature in presence of a catalyst to give nitric oxide, which further combines with oxygen to give nitrogen dioxide. The dissolution of nitrogen dioxide into water gives nitric acid. This is the basis of Oswald's process for the manufacture of nitric acid. The nitric acid, thus obtained is converted into nitrates, which are used as fertilizers.

The flow chart for the fixation of nitrogen and the various features in nitrogen cycle are given in Figs 1.6 and 1.7, respectively.

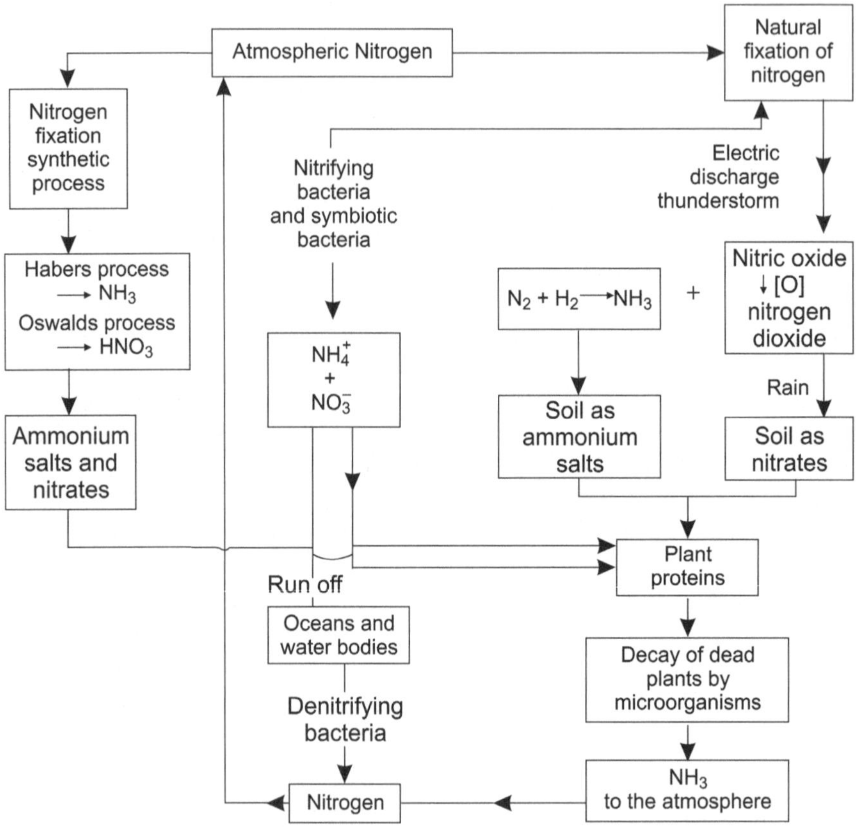

Fig. 1.6 Flow chart for nitrogen fixation

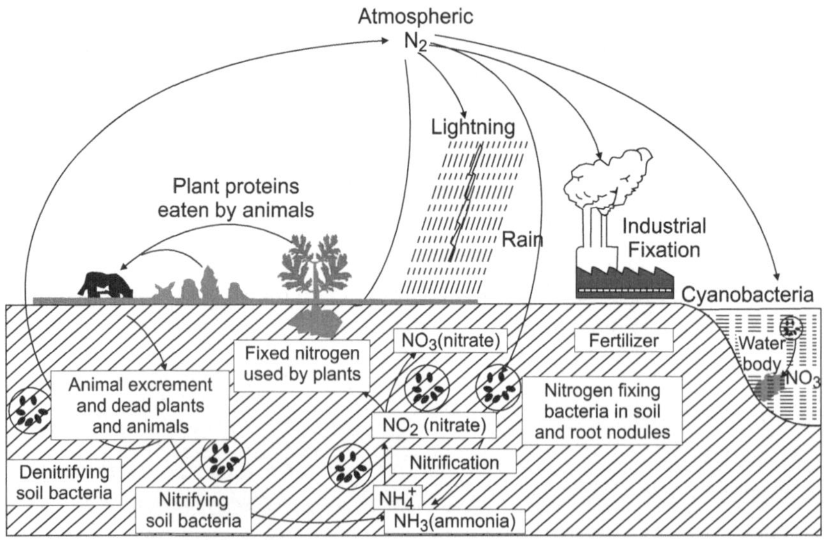

Fig. 1.7 Nitrogen cycle-various features

1.8.3 Oxygen Cycle

Oxygen is the second most abundant element in the atmosphere and is present to the extent of about 21 per cent and is essential for all life processes. Atmosphere and oceans are the main pool for oxygen. The oxygen dissolved in water (oceans, etc.) is responsible for sustaining aquatic plants and animals. It is also a constituent of oxide and carbonate ores. There is a continuous exchange of oxygen between the atmosphere and water surfaces of all water bodies. The total amount of oxygen in the biosphere is constant. The oxygen cycle is based on the exchange of oxygen among the segments of the environment, viz., atmosphere, hydrosphere, lithosphere and biosphere. It plays a key role in atmospheric chemistry and life process. The different steps involved in oxygen cycle are:

- During respiration, organisms including animals and human beings consume oxygen from the air and give carbon dioxide into the atmosphere.
- The green plants utilise carbon dioxide for photosynthesis, in which carbon dioxide reacts with water in presence of solar energy and a catalyst (chlorophyll) to produce carbohydrates and the formed oxygen is released into the atmosphere.

$$6CO_2 + 6H_2O + \text{Solar energy} \xrightarrow[\text{Chlorophyll}]{\text{Photosynthesis}} C_6H_{12}O_6 + 6O_2$$

- Oxygen is needed by bacteria and fungi during decomposition of organic matter. The products of decomposition (CO_2 and H_2O) are released into the atmosphere. The released CO_2 and H_2O are utilised during photosynthesis by green plants.
- In the upper atmosphere, solar energy splits H_2O to give H_2 and O_2, which are released into the atmosphere

$$2H_2O + h\nu \longrightarrow 2H_2 + O_2$$

The above processes complete the oxygen cylce. The oxygen content in the atmosphere remains more or less constant. The oxygen cycle is represented as shown in Fig. 1.8.

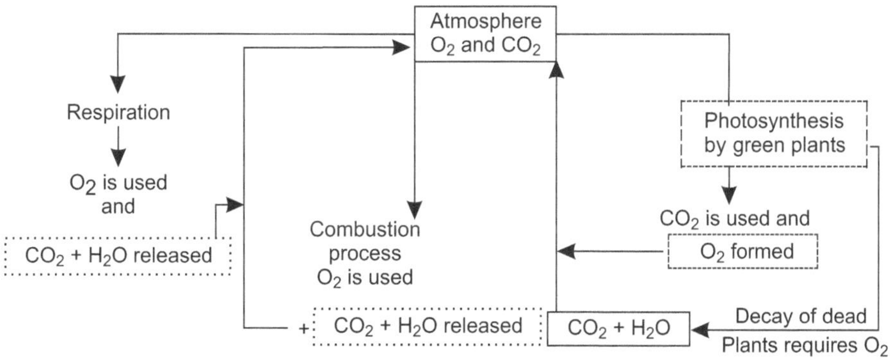

Fig. 1.8 Oxygen cycle flow chart

Under certain circumstances, there may be imbalance in the oxygen cycle. For example, deforestation reduces the extent of photosynthesis (which is a major source of oxygen in the atmosphere); this leads to reduction of oxygen content in the atmosphere. This loss, however, is compensated by sulphate reducing bacteria occurring in anaerobic environment; these bacteria use sulphate ions as an oxygen source for oxidizing organic matter.

$$SO_4^{2-} + 2C \longrightarrow 2CO_2 + S^{2-}$$

The above reaction gives carbon dioxide, which is released into the atmosphere and utilized in photosynthesis leading to the release of oxygen.

1.8.4 Phosphorus Cycle

Phosphorus plays a vital and indispensable role in living organisms. Phosphorus in the form of phosphates are necessary for the growth and maintenance of animal bones and teeth. It is one of the most important element required by plants for their growth. It is a constituent of nucleic acids, phospholipids, ATP, and ADP. The main reservoirs of phosphorus are rocks and other deposits. From these, phosphorus is made available to the living organisms via a slow process of weathering of the rocks. Phosphorus is absorbed by the plants through their root system, in its oxidized form, as phosphate, where it gets incorporated into different phosphorus, containing compounds. From plants, phosphorus finds its way to other living organisms. The decay of dead organisms and plants release the phosphorus content back to the soil, which is again recycled. Some phosphorus finds its way to the water bodies (rivers and oceans). In oceans, the phosphorus is consumed by fish and sea birds and any excess phosphorus in deposited into the sediments. The flow chart of phosphorus cycle is represented in Fig. 1.9.

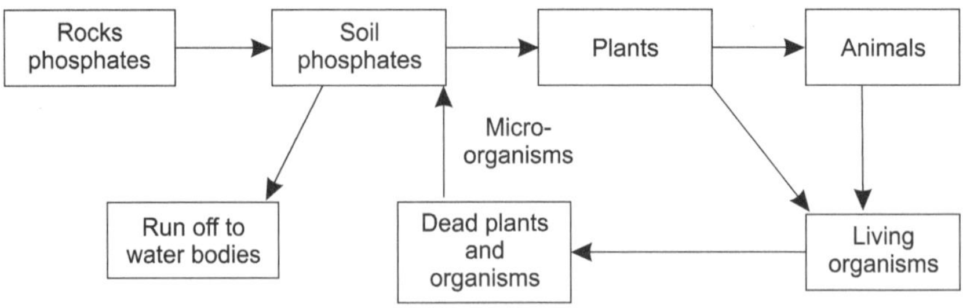

Fig. 1.9 Flow chart of phosphorus cycle

In fresh water, the floating algae (phytoplankton) absorb soluble inorganic phosphate and convert them into organophosphates. The algae are eaten by zooplankton, which in turn are eaten by other aquatic animals like fish. The aquatic animals after their death and decay by microorganisms settle to the bottom of water. Finally the microorganisms release phosphates into the water bodies and the cycle continuous (Fig 1.10).

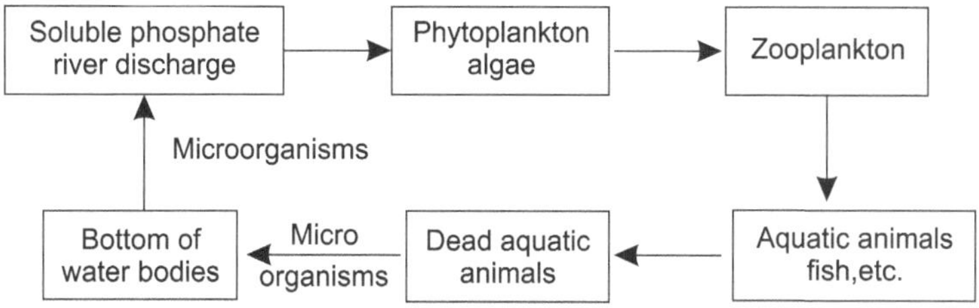

Fig. 1.10 Flow chart of phosphorus cycle in water bodies

A major problem associated with the presence of excess of phosphate fertilizers (and also other fertilizers) is eutrophication of lakes. The increase in the nutrients (by agricultural run offs) in the water bodies is called **eutrophication**. These nutrients, particularly phosphorus cause population explosion of photosynthetic bacteria and blue-green algae. With the passege of time, the whole of the lake or pond is covered by a thick layer of algae resulting in depletion of the availability of oxygen. This results in the death of algae, bacteria and fish.

1.8.5 Sulphur Cycle

The main source of sulphur is the lithosphere, where it is present as free sulphur and as sulphide ores of different elements like iron, copper, lead, tin, mercury, zinc, etc. Sulphur is also present in coal and oil in the sedimentary rocks. Sulphur is a constituent of two of the twenty essential amino acids and is incorporated in several proteins and living organisms. The weathering of sedimentary rocks releases sulphur. Decomposition of organic matter by bacteria and fungi is also responsible for releasing sulphur as hydrogen sulphide. In the gaseous phase, sulphur is present as hydrogen sulphide and sulphur dioxide. Sulphur as sulphur dioxide enters the atmosphere from several routes like volcanic eruptions, combustion of fossil fuels and from various metallurgical operations. Hydrogen sulphide enters the atmosphere by the putrification of organic matter by bacteria and fungi.

In the atmosphere, SO_2 in oxidised to SO_3 which comes to the land in the form of dilute solution of H_2SO_4 (acid rain). Hydrogen sulphide in the atmosphere is oxidised to sulphur dioxide by ozone; the reaction is catalysed by particulates present in the atmosphere.

$$H_2S + O_3 \longrightarrow H_2O + SO_2$$

In anaerobic soils and sediments, hydrogen sulphide is formed by sulphate reducing bacteria like *Desulfavibrio*. Certain species of *Beggiat* oxidise hydrogen sulphide to elemental sulphur. Even green and purple sulphur photosynthetic bacteria oxidise hydrogen sulphide to elemental sulphur. The flow chart of sulphur cycle is represented in Fig 1.11.

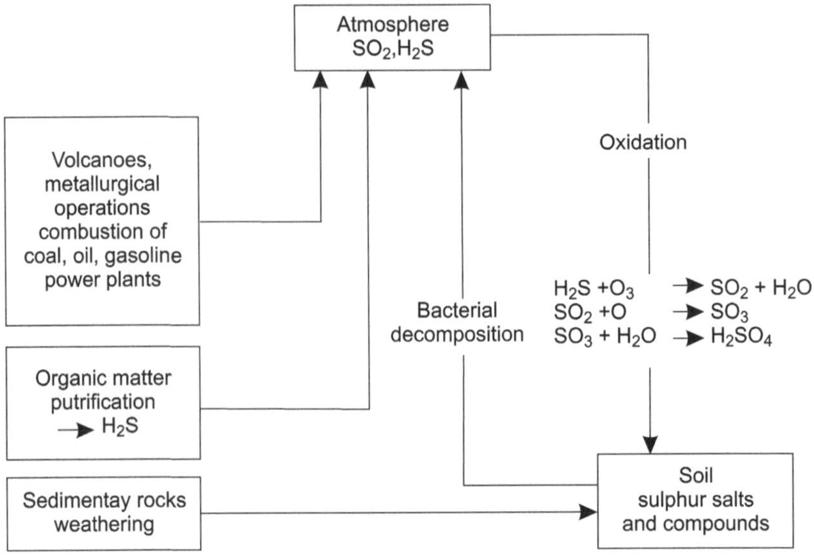

Fig. 1.11 Sulphur cycle (flow chart)

1.9 BIODISTRIBUTION OF ELEMENTS

For biological functions, the living organisms require 27 elements of which 15 are metals. The elements required in major quantities are potassium, magnesium, sodium and calcium. Minor quantities of manganese, iron, cobalt, copper, zinc and molybdenum and trace amounts of vanadium, chromium, tin, nickel and aluminium are required at least by some organisms.

Bulk quantities of group 1 and group 2 metals are required to balance the electrical charges associated with negatively charged organic macromolecules in the cell, and also to maintain the osmotic pressure inside the cell, to keep it turgid and prevent its collapse.

The elements required for living organisms are found in earth's crust, rivers and sea water. Table 1.4 gives the abundance of the elements in earth's crust, river water and ocean water.

Table 1.4 Abundance of Some Elements in Land and Water

Elements	*Earth's Crust*		*River Water*	*Ocean Water*
	g kg^{-1}	*atom/10^4 atoms si*	*mg L^{-1}*	*mg L^{-1}*
Potassium	24	610	2.3	416
Mangnesium	20	820	4.1	1.326×103
Sodium	24	1040	9	1.05×104
Calcium	42	1040	1.5	4.22

(Table 1.4 Contd...)

Elements	***Earth's Crust***		***River Water***	***Ocean Water***
	g kg^{-1}	***atom/10^4 atoms si***	***mg L^{-1}***	***mg L^{-1}***
Maganese	0.95	17	~0.005	0.002a
Iron	56	1000	0.67	0.003a
Cobalt	0.025	0.42	0.0002	8×10^{-5a}
Copper	0.055	0.86	0.005	0.003[a]
Zinc	0.070	1.1	0.01	0.005
Molybdenum	0.0015	0.016	0.001	0.01
Vanadium	0.135	2.64	0.001	0.0015
Chromium	0.1	2	0.001	0.0006[a]
Tin	0.002	0.02	4×10^{-5}	1×10^{-5}
Nickel	0.075	1.3	0.0003	0.002
Aluminium	82	3020	0.4	0.005[a]

[a]Considerable variation occurs.

The data in the above table is taken from Brown, H.J.M. Environmental Chemistry of Elements, Academic, New York 1979; Chapter 13.

Different elements have different role to play in biological systems. Table 1.5 give the functions and toxicity some of the elements in biological systems.

Table 1.5 Function and Toxicity of Some of the Elements In Biological Systems

Element	***Biological Function***	***Toxicity***	***Comments***
Potassium	Essential to all organisms with the possible exception of blue-green algae; major cation in intracellular fluids in animals; essential for transmission of nerve impulse and cardiac function.	Extremely toxic to mammals when injected intravenously; emesis prevents oral toxicity.	Pollution problem possible from leaching of fertilizers from agricultural land.
Magnesium[1]	Essential to all organisms. Present in all chlorophylls. Has other electrochemical and enzyme-activating functions.		May cause deficiencies of other elements (e.g. Fe) by the effect of the alkalinity of dolomite.

(Table 1.5 Contd...)

Element	Biological Function	Toxicity	Comments
Sodium	Important for nerve functioning in animals. Major cation of extracellular fluid in animals.	Relatively harmless except in excessive amounts (lethal dose ca. 3 g kg^{-1}). Associated with some forms of hypertension.	Tolerance of and/or dependence upon sodium chloride can be an important consideration for the survival of plants and aquatic animals. This depends upon osmotic regulation rather than sodium specificity.
Calcium[1]	Essential for all organisms; used in cell walls, bones, and some shells as structural component; important electrochemically and involved in blood clotting.	Relatively harmless.	May cause deficiencies of other elements (e.g. Fe) by effect of alkalinity of lime stone.
Manganese[2]	Essential for all organisms; activates numerous enzymes; deficiencies in soils lead to infertility in mammals, bone malformation in growing chicks.	Moderately toxic.	
Iron[3]	Essential to all organisms.	Normally only slight toxicity, but excessive intake can cause siderosis and damage to organs through excessive iron storage (hermochromatosis).	A very abundant element (5 per cent of earth's crust); may not be available at high pHs.
Cobalt[4]	Essential for many organisms including mammals; activates a number of enzymes; vitamin B_{12}	Very toxic to plants and moderately so when injected intravenously in mammals.	Extensive areas are known where low soil cobalt affects the health of grazing animals.
Copper[5]	Essential to all organisms; constituent of redox enzymes and hemocyanin.	Very toxic to most plants; highly toxic to invertebrates, moderately so to mammals.	Pollution from industrial smoke and possibly from agricultural use. Wilson's disease, genetic recessive, results in toxic increase in copper storage.

(Table 1.5 Contd...)

Element	Biological Function	Toxicity	Comments
Zinc[6]	Essential to all organisms; used in >70 enzymes; stabilizes coiled ribosomes. Plays a role in sexual maturation and reproduction.	Moderately to slightly toxic; orally causes vomiting and diarrohea.	Pollution from industrial smoke may cause lung disease: use of zinc promotes cadmium pollution. Certain areas (e.g. Iran and Egypt) are zinc deficient.
Molyb-denum[7]	Essential to all organisms with the possible exception of green algae; used in enzymes connected with nitrogen fixation and nitrate reduction.	Moderatley toxic and antagonistic to copper—molybdenum excesses in pasturage can cause copper deficiency.	Pollution from industrial smoke may be linked with lung disease.
Vanadium	Essential to ascidians ('sea squirts'), which concentrate in a million-fold from sea water. Essential to chicks and rats. Deficiencies cause reduced growth, impaired reproduction and survival of young, impaired tooth and bone metabolism and feather development. May be a factor in manic— depressive illness.	Highly toxic to mammals if injected intravenously.	Possible pollutant from industrial smokes—may cause lung disease.
Chromium[7]	Essential; involved in glucose metabolism and diabetes; potentiates effect of insulin. Presence in glucose tolerance factor from brewer's yeast questioned.	Highly toxic as Cr(VI); carcingenic; moderately toxic as Cr(III).	Potential pollutant since amount used industrially is large compared with normal biological levels; normally relatively unavailable because of low solubility. Cr(VI) used in comfort cooling towers, environmental hazards.

(Table 1.5 Contd...)

Element	Biological Function	Toxicity	Comments
Tin	Weak evidence for ultratrace essentiality in rats.	Organotin compounds used as bacteriostats and fungistats; its use in anti-foulant boat paints now discouraged because of danger to estuarine and marine life.	
Nickel	Essential trace element. Chicks and rats raised on deficient diet show impared liver function and morphology;[c] stabilizes coiled ribosomes. Active metal in several hydrogenases and plant ureases.[e]	Very toxic to most plants, moderately so to mammals; carcinogenic.	Local industrial pollutant of air and water.
Aluminium[8]	May activate succinic dehydrogenase and δ-aminolevulinate dehydrase. The latter is involved in porphyrin synthesis.	Moderately toxic to most plants; slightly toxic to mammals. Suggested as involved in the etiology of Alzheimer's disease and other neurologic diseases.	Relatively inaccessible except in acidic media as a result of insolubility of $Al(OH)_3$. Soils and waters high in Al^{3+} and low in Mg^{2+} and Ca^{2+} implicated in neurologic diseases.

***(i)* Biological role of Mg^{2+} and Ca^{2+}.** Mg^{2+} ions are concentrated in animal cells and Ca^{2+} ions are concentrated in body fluids outside the cell, in much the same way that K^+ concentrates inside the cell and Na^+ outside. Mg^{2+} ions form a complex with ATP and are constituents of phosphohydrolases and phosphotranferases, which are enzymes for reactions involving ATP and energy release. They are also essential for transmission of impulses along the nerve fibres. Mg^{2+} is important in chlorophyll, in the green plants. Ca^{2+} is important in bones and teeth as apatite, $Ca_3(PO_4)_2$, and the enamel on teeth as fluoroapatite, $[3(Ca_3(PO_4)_2).CaF_2]$. Ca^{2+} ions are important in blood clotting and are required to trigger the contraction of muscles and to maintain regular beating of the heart.

***(ii)* Biological importance of manganese.** Manganese is biologically important and is necessary for photosynthesis. It is important in both animal and plant enzymes. The importance is because it converts nitrogenous waste products into urea in the ornithine-arginine-citrilline cycle (discussed by Hans Krebs who discovered the tricarboxylic cycle). The urea is carried by the blood to the kidneys, where it is excreted as urine.

Manganese is also an essential trace element for plant growth. It is added to the fertilizers in parts of the world where there is deficiency in the soil. It is essential in a group of enzymes called phosphotransferases.

***(iii)* Iron.** Iron is essential in small amounts for both plant and animal life. However, like copper and selenium, it is toxic in larger quantities. Biologically iron is the most important transition element. It is involved in several different processes.

- As an oxygen carrier in the blood of mammals, birds and fish (haemoglobin)
- For oxygen storage in muscle tissues (mycoglobin)
- As nitrogenase (the enzyme in dinitrogen fixing bacteria)
- As a number of other enzymes: aldehyde oxidase (oxidation of aldehydes), catalase and peroxidase (decomposition of H_2O_2) and succinic dehydrogenate (the aerobic oxidation of anhydrides).

***(iv)* Cobalt.** Cobalt is also biologically important in some enzymes. Glutamic mutase is involved in the metabolism of amino acids and ribonudeotide reductase in the biosynthesis of DNA. Traces of cobalt are essential in the diet of animals. Some sheep raised in Australia, New Zealand, Florida and Britain suffer from a deficiency disease which was traced to their grazing on cobalt deficient diet soil. This can be remedied either by treating the soil periodically or by forcing the animal to swallow a pellet of cobalt. Large amount of cobalt appear to be harmful.

***(v)* Copper.** Copper is biologically important in various oxidase enzymes, an oxygen carrier in invertebrates and in photosynthesis. Copper is essential to life and adult human contain about 100 mg. Though small amounts of copper are essential, larger amounts are toxic. About 4–5 mg of copper are required daily in the diet, and deficiency in animals result in the inability to use iron stored in the liver. Thus animals suffer from anaemia. The copper is bound to proteins in the body either as metalloproteins or as enzymes. Examples include various oxidases and blue proteins. These include Amine oxidase, ascorbate oxidase, cytochrome oxidase and galactose oxidase.

Copper is also important in Lysine oxidase (which affects the elasticity of aortic walls), Dopamine hydroxylate (which affects brain function), Tyrosine (which affects skin pigmentation) and ceruloplasmin (which plays a role in metabolism).

Wilsons disease is a hereditary shortage of ceruloplasmin, causing copper accumulation in the liver, kidneys and brain. This distance can be treated by feeding the patient with EDTA. The copper forms a complex which is excreted.

***(vi)* Zinc.** Zinc plays an important role in the enzyme systems of animals and plants. Humans contain about 2 g. of zinc. This is the second largest amount of a transition metal after iron and molybdnium. There are about twenty enzymes containing zinc. Some of the best known are:

- **Carbonic anhydrase.** It is present in red blood cells and is involved in respiration. It speeds up the absorption of CO_2 by red blood cells in muscles and other tissues, and the reverse reaction involving the release of CO_2 in the lungs. At the same time, it regulates the pH.

$$CO_2 + OH^- \rightleftharpoons HCO^-_3$$

- **Alkaline phosphate** (energy release)
- **Dehydrogenases** and **aldolases** (sugar metabolism)
- **Alcohol dehydrogenase** (metabolism of alcohol)
- **Carboxypeptidase.** This zinc containing enzyme is particularly useful for animals in the digestion problem. It works by catalysing the hydrolysis of terminal peptide link at the end of the peptide chain.

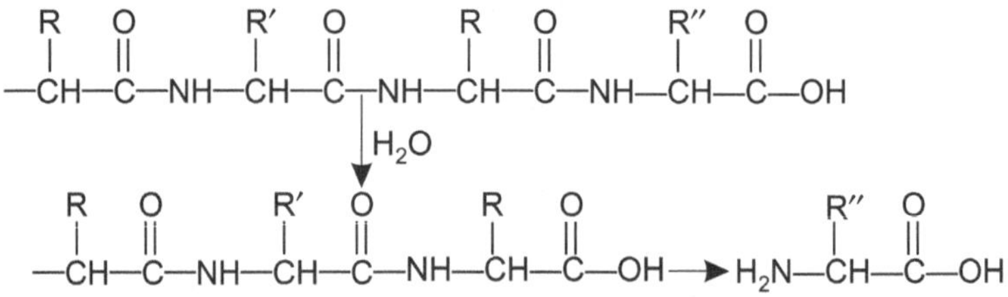

The enzyme carboxypeptidase is found to be present in pencretic juice.

***(vii)* Biological importance of chromium and molybdenum.** Trace amounts of chromium and molybdnium are necessary in the diet of mammals. Cr (+ III) and insulin both are involved in maintaining the correct level of glucose in the blood. In case of chromium deficiency, glucose is only removed from the blood half as fast as normally. Some cases of diabetes may reflect faulty metabolism of chromium. The most important medical aspect of chromium salts is that larger amounts either ingested or on the skin are carcinogenic.

Molybdenum is present in the catalysts of nitrogen fixing bacteria. The amount of dinitrogen fixed biologically is estimated at 175 million ions per year compared with a total of 110 million tons of NH_3 produced by the Haber-Bosch process and distillation of coal. The best known dinitrogen fixing bacteria is Rhizobium. This contains the metalloenzyme nitrogenase. Nitrogenase contains two proteins, molybdoferrodoxin and azoferedoxin. The former is air sensitive and contains two Mo atoms, 24–36 Fe atoms and 24–36 S atoms together with a protein and has a molecular mass of about 225000.

***(viii)* Aluminium.** It is a silvery white metal and belongs to group 13 to the periodic table. It occurs abundantly in earth's crust; the important ores are bauxite cond cryolite. Tradiationally, aluminium has been used for making utensils for cooking purposes. However, it is believed that aluminium is not as harmless as it was believed to be. Some of the salts of aluminium are used for treatment. A typical example is aluminium hydroxide, which is used as an anti-acid in case of indigestion. Water is purified by adding aluminium sulphate.

The element aluminium is toxic but is usually excreted from the system. However, people with kindly failure are unable to excrete aluminium and so because toxic. In case aluminium salts find their way in the brain, it is believed to cause Alzheimers disease.

Most of the elements from the soil are taken up by the plants and find their way into animals and humans.

EXERCISE

1. What do you understand by environmental chemistry? What is its importance?
2. What are the components of the environment?
3. Describe the segments of the atmosphere.
4. What is the importance of Biosphere?
5. What are the major components of the atmosphere?
6. What are the main constituents of the different regions of the atmosphere?
7. Carbon dioxide is not poisonous but a person left in CO_2 atmosphere dies, why?
8. What do you understand by heat budget of the earth?
9. What do you understand by biogeochemical cycles?
10. Write notes on:
    (*i*) Carbon cycle
    (*ii*) Nitrogen cycle
    (*iii*) Oxygen cycle
    (*iv*) Phosphorus cycle
    (*v*) Sulphur cycle
11. Write a note on biodistribution of elements.
12. Explain the biological role of Mg^{2+}, Ca^{2+}, Fe, Zn, Cr and Mo.

❑❑❑

Chapter 2

Hydrosphere

2.1 INTRODUCTION

A hydrosphere (Greek *hydro* means 'water') in physical geography described the collective mass of water found on, under, and over the surface of a planet. Hydrosphere covers more than two-thirds of the earth's surface either as ocean (salt water) or as fresh water, which includes seas, rivers, lakes, ponds, etc. The oceans account for about 95 per cent of earth's water supply. This is unfit for consumption due to its high salt content. The polar ice caps and glaciers account for about two per cent of the remaining water. Fresh potable water is available only to the extent of about one per cent in rivers, lakes and as groundwater.

In hydrosphere, water is present in all the three forms, viz., solid, liquid and gaseous. The major portion of hydrosphere are the oceans. Besides oceans, water is also present in inland seas, lakes, rivers, underground water, water vapour in the atmosphere and glaciers (in which water is present as snow) on an average, the depth of oceans is about 3, 794 m.

The hydrosphere is called the 'Blue planet'; this is due to the looking at sea, which looks blue. In hydrosphere, approximately 70.8 per cent of earth is covered by water. Out of this water, about 97 per cent is the sea water. The availability of fresh water in the hydrosphere is only 3 per cent.

Water is the most essential component of life. It is unique liquid in the sense that without it no life is possible. Though on a global scale, there is abundance of water, but the main problem is the availability of water in the right place at the right time and in the right from. The survival of mankind has always depended on water. It is well known that in earlier times, civilizations flourished on river beds. Also, the river beds were responsible for whiping out the populations due to floods.

Water can be found in liquid, solid or gaseous form in a number of places. Depending on the specific location, the residence time may vary from a few days to thousands of years (Table 2.1).

Table 2.1 Global Water Supply

Location source	*Percentage of total water*	*Approximate average residence time*
Oceans	97.2	Infinity
Atmosphere	0.001	9 days
Rivers and streams	0.0001	2 weeks
Ground water (up to a depths of 0.8 km)	0.31	Hundreds to thousands of years
Lakes (fresh water)	0.009	Tens of years
Ice caps and glaciers	2.15	Tens of thousands of years

Source : US Geological Survey.

Water has a high value of latent heat and it is this property that moderates the temperature of the biospheres. It is a universal solvent. The solubility of oxygen in water is responsible for survival of aquatic life. Though there is loss of water by evaporation, but its amount remains constant. This is because there is a continuous circulation of water including water bodies and natural processes which use water. Due to rapid industrialization, human beings are responsible for the discharge of waste materials into water bodies like seas, oceans, rivers, etc. This has created serious problems for the survival of life not only on the land but also the survival of aquatic life.

2.2 CHARACTERISTICS OF WATER

Water is a unique liquid, without which no life can survive. Water is a liquid at room temperature, with mp. 0°C, bp. 100°C, high dipole moment (1.84×10^{-3} esu), high dielectric constant (80), density (1.0), specific heat (1 Cal g^{-1} $°C^{-1}$), high heat of evaporation (540 Cal g^{-1}), surface tension (73 dynes cm^{-1} at 20°C) and viscosity (0.01 poise at 20°C). These abnormal properties of water are due to hydrogen bonding. The uniqueness of water is attributed to the following properties.

(*i*) Water is regarded as a **universal solvent**. No other liquid can be compared to water as a solvent. Natural water is slightly acidic (pH 5.6) due to dissolved carbon dioxide, which gives carbonic acid (H_2CO_3). Therefore, it can dissolve a variety of compounds from simple salts to minerals including sodium chloride and calcium carbonate in limestone rock. Water also reacts with complex organic compounds including many amino acids found in the human body.

(*ii*) Water possesses highest heat of fusion and heat of evaporation, collectively known as latent heat, of all known liquid substances at ordinary temperature. The latent heat of water is responsible for moderating the temperature of the biosphere. It also plays a vital role in the evaporation of water and its condensation as rain and as dew (see hydrological cycle).

(*iii*) Compared with other common liquids, water has a high surface tension, a characteristic that is very important in many physical and biological processes involving movement of water through and into organisms.

(*iv*) Water is a poor thermal conductor as compared to other metals.

(*v*) The high viscosity of water allows organisms to swim using simple movements.

(*vi*) Water protects aquatic animals and plants from mechanical disturbances.

(*vii*) Among common compounds and molecules, water is the only one whose solid form (ice) is lighter than its liquid form; it expands by about 8 per cent when it freezes and becomes less dense. This is responsible for the floating of ice on water. Can you imagine, if ice were heavier than water, it would sink to the bottom of water bodies; in such a situation, the biosphere would be entirely different than what it is today and life if existed could be greatly altered. Freezing of water bodies from the bottom onwards will result in complete freezing of water bodies resulting in the death of all life in water.

(*viii*) Being transparent to light, water allows photosynthetic organisms to live below the surface of water.

(*ix*) Water exerts pressure. Organisms living at the sea level experience a pressure of about 15 psi (~1 atmosphere, which is equal to 360 mm mercury). With depth, the pressure increases at the ratio of about one atmosphere for every 10 meters of decent. The physical characteristics like solubility, ionic dissociation and surface tension are all influenced by the pressure.

2.3 CHEMICAL COMPOSITION OF WATER BODIES

The chemical composition of water bodies, e.g. ground water, surface water (river, lakes), oceans, etc., depends on their source.

2.3.1 Ground Water

Water is a gift of nature. Over 98 per cent of the fresh water on the earth lies below the earth as ground water. It is the main source of drinking water for about 50 per cent of the global population.

It is generally believed that ground water is pure and safe to drink. However, this is not true. In fact ground water can easily be polluted from wastes generated from domestic, industrial and agricultural sources. The average composition of ground water is given in Table 2.2.

Table 2.2 Chemical Composition of Ground Water

Parameters	*Amount (ppm)*
Sodium	8.2
Potassium	1.4
Calcium	92
Magnesium	34
Iron (III)	0.09
Silica	10
Chloride	9.6
Bicarbonate	330
Nitrate	13
Sulphate	84
Total dissolved solid	434

2.3.2 River Water and Lake Water

River water and lake water are referred to as surface water. The rivers, as we know originate from the mountains and on way dissolve various minerals they come across. The lakes are formed mostly from river water.

The river water originally is pure, but on its way it gets polluted from surface run off, industrial and municipal waste, agricultural wastes, decomposed plant and animal matter and radioactive materials. The average composition of surface water is given in Table 2.3.

Table 2.3 Chemical Composition of River Water

Parameters	*Amount (ppm)*
Sodium	6.5
Potassium	1.2
Calcium	36
Magnesium	8.0
Iron (III)	0.02
Silica	1.2
Chloride	13
Bicarbonate	119
Nitrate	0.1
Sulphate	22
Total dissolved solid	1.65

2.3.3 Sea Water

The oceans have been used by man for centuries for recovering salt and other chemicals. In recent years, besides the chemicals, the oceans have become a good source of petroleum oil, natural gas, etc. The oceans get their water from various rivers. Since there is continuous evaporation from sea, the concentration of various constituents goes on increasing. The chemical composition of sea water is given in Table 2.4.

Table 2.4 Chemical Composition of Sea Water

Parameters	*Amount (ppm)*
Major constituents	
Sodium	10561
Magnesium	1270
Sulphur	880
Calcium	400
Potassium	380
Chloricle	18980
Minor constituents	
Bromine	65
Carbon	28
Strontium	13
Boron	4
Trace constituents[1]	
Lithium	0.1
Iodine	0.05
Copper	0.003
Silver	0.0003
Gold	0.000006

[1] Besides the trace constituents given in the table, sea water contain about 25 other trace constituents in very minute amounts. These include Rb, Ba, Zn, Al, Mo, U, Ni, Mn, V, Ti, Co, Cd, Cr, Hg, Pb, F.

The composition of the ocean is generally expressed in terms of **salinity**. Salinity is defined as the total amount in grams of solid material dissolved in one kilogram of sea water, when all the carbonate has been converted to oxide, all the iodide and bromide have been replaced by chloride, and the organic matter has been completely oxidised. Salinity is expressed as gram of salt per kilogram of sea water (g kg^{-1}), that is, as parts per thousand by weight (%). Table 2.5 gives the contribution to salinity of various inorganic salts. The salinity of all the oceans and seas is about 35 per cent. However small variations in salinity have been observed from place to place.

Table 2.5 Main Salts in the Ocean

Salt	%
Sodium Chloride	23.0
Magnesium Chloride	5.0
Sodium Sulphate	4.0
Calcium Chloride	1.0
Potassium Chloride	0.7
Others	1.3
Total	35.0

2.3.4 Wetlands

Wetlands are shallow water zones adjacent to aquatic regions (river, sea etc). In wetlands, water is near its surface. Wetlands are effective in flood control. They act as a buffer against the effects of hurricanes and cyclones. They check soil erosion by sea. They are a source of durable timber, fuel wood, fruits, vegetables, traditional medicines and protein rich fodder for cattle. Thus, wetlands are of great economic and scientific value. In India, there are 27 wetlands which are present in 15 states.

The chemical composition of the waters in wetlands depend on the source of water, whether derived from rivers or oceans or both.

2.4 HYDROLOGICAL CYCLE

Hydrological cycle, also known as water cycle is the most important of the various cycles of the environment. It involves natural process resulting in the exchange to water between various water bodies, i.e., the oceans, seas, rivers etc. The hydrological cycle involves three different and continuous processes, i.e., evaporation of water followed by condensation leading to rainfall. Finally the rain water returns to the water bodies. Water also finds its way into ground water by seepage from the land. The ground water is again taken up by plants, which in turn release water as vapours in the atmosphere. The hydrological cycle can be compared to distillation, which involves vaporisation of water by heat. The vapours on cooling get converted back into liquid form.

The water in the oceans moves as it is of different temperature and salinity at different locations. Surface waters are also moved by winds, giving rise to surface ocean currents. Warm water is lighter or less dense than cold water which is more dense or heavier and salty water is also more dense than fresh water. The combination of the water's temperature and salinity determines whether it rises to the surface, sinks to the bottom, or stays at some intermediate depth.

Table 2.6 Global Distribution of Water

Source	*Amount in percentage*
Oceans	97.6
Ice in polar regions and glaciers	1.8699
Ground water	0.5
Rivers, lakes, inland seas	0.02
Soil moisture	0.01
Atmosphere	0.0001

Water circulates between the living and non-living components of the biosphere in the form of unending cycles. (Fig. 2.1). In hydrological cycle, there is movement of water from the oceans to the atmosphere by evaporation and from atmosphere to oceans and land by precipitation, from land to oceans by run off and from streams and rivers and from land to the atmosphere by evaporation. The energy required for the above cycle is provided by the sun in the form of solar energy. In the above cycle, water in oceans, seas and lakes, etc. circulates between them and the atmosphere.

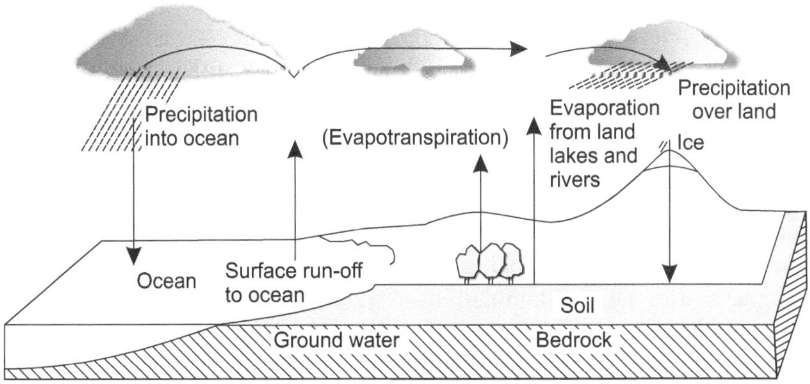

Fig. 2.1 The movement of water around, over, and through the Earth is called the **water cycle**, a key process of the hydrosphere

In an alternative pathway for hydrological cycle, the soil water or the underground water is circulated by plants, animals and the atmosphere. The plants take up the soil water through their root system. A part of this water is utilised by plants for photosynthesis and the excess water is given off into the atmosphere during transpiration. (Transpiration is the loss of water from the leaves of the plants via evaporation). Transpiration, in fact, is the main source of water into the atmosphere. The decay of dead plants by microorganisms release water back into the soil. Animals consume water by drinking from various sources and also via plants they eat. Water vapour is again released into the atmosphere by breathing and evaporation from the surface of animals (sweating) and excretion. Whatever the route is followed, the water vapours being lighter rise in the air, where it gets cooled and condenses into tiny droplets to form a cloud. The water from the clouds ultimately falls on the earth in the form of rain, part of which gets absorbed by the soil and the remaining finds its way into water bodies. (see Fig. 2.2).

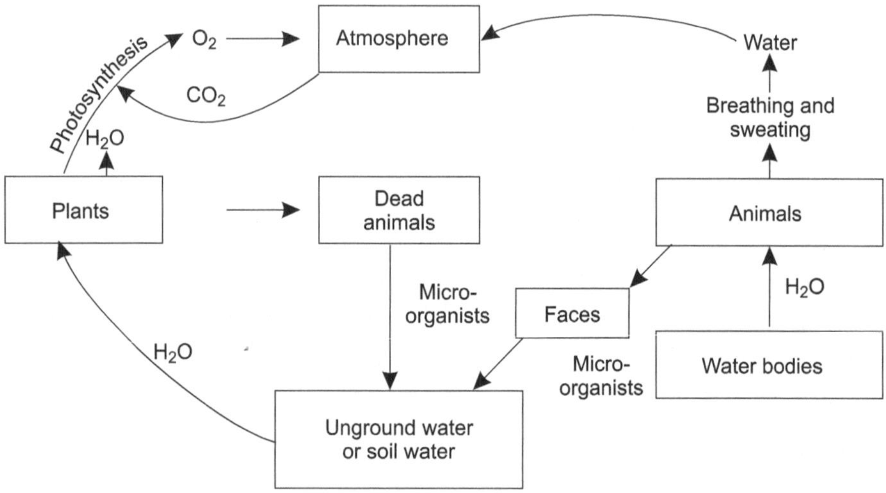

Fig. 2.2 Water cycle (alternative pathway) flow chart

2.5 AQUATIC POLLUTION

2.5.1 Introduction

Aquatic pollution or water pollution, as it is commonly called, is the degradation of the quality of water that makes it unsafe or harmful to human beings, animals and aquatic life. The contamination of water, to a certain extent is inevitable, since water is a universal solvent and can dissolve various types of substances in it. Aquatic pollution disturbs the normal use of water for irrigation, agriculture, industries and for human consumption. Aquatic pollution can be caused by natural or anthropogenic processes. Natural processes include pollution of water bodies by decomposed vegetables, animals and weathered products. The anthropogenic process include industrial, radioactive, mining processes.

2.5.2 Types of Water Pollutants

Water pollutants include biological agents, chemical agents and physical agents.

(*i*) Biological Agents: These include bacteria, parasitic fungi and protozoa. The main sources include human sewage, animal and plant wastes, decaying organic matter, industrial wastes (oil refineries, oil spills, oil pollutants, paper mills, food processing units, etc.) and natural land and urban run off and waste water discharges from slaughter houses. Some of the contributory factors are over population, unplanned industrial and human settlements and lack of civic amenities.

The oxygen requiring bacteria feed on the biological wastes and this results in depletion of oxygen in the water bodies. In the absence of oxygen, life is destroyed. The water acquires foul odour.

***(ii)* Chemical Agents:** These can be inorganic chemicals and minerals, organic chemicals and radioactive substances. The **inorganic chemical agents** include acids, salts (nitrates, phosphates etc.), metals like lead and mercury. The main sources include natural run off from land, industrial wastes, leaded gasoline, lead smelting, agricultural chemicals run off from agricultural lands, mining, oil fields, food processing units, etc. In water bodies, the above pollutants are toxic to various forms of life and human beings and can cause genetic defects. The water is rendered unfit for human consumption (domestic, agricultural and industrial use). It is responsible for eutrophication.

The **organic chemicals** include agrochemicals (pesticides, herbicides etc), detergents, chlorine containing compounds, oil, grease and plastics. The main source include agriculture, forestry, pest control industries, domestic and industrial wastes, paper industry, bleaching processes etc. Such polluted water is toxic to aquatic life and other organisms that are dependent of such water bodies. It may cause eutrophication of water bodies.

The **radioactive** substances are released into aquatic systems as a result of nuclear wastes from research laboratories and hospitals. These are also released by processing of uranium ores and nuclear plants. The radioactive substances enter the food chain and are responsible for birth and genetic defects. They are causative agents for cancer.

***(iii)* Physical agents:** These include suspended solids, particulates and heat. The main sources are soil erosion, run off from agriculture, mining and power plants. Suspended organic and mineral solids absorb toxic substances like heavy metals and pass them through food chain. Aquatic animals face problems in breathing through gills in such waters. Thermal pollution occurs when hot water enters the water body. Increase in temperature lowers the solubility of oxygen in water thereby affecting aquatic life.

2.5.3 Types of Aquatic Pollution

Depending on the source and storage of water aquatic pollution can be of three types. These are ground water pollution, surface water pollution and sea water or marine pollution.

Ground Water Pollution

Ground water gets polluted due to human activities (e.g. industrial, domestic and agricultural activities). Normally, ground water pollution is irreversible. This leads to the degradation of the water quality, i.e. objectionable taste, odour and excessive hardness. Following Table 2.7 gives the sources of contamination of ground water.

Table 2.7 Sources of Contamination of Ground Water

Sources of contamination	*Contaminates discharged in ground water*
Domestic Waste	Pathogenic organisms, nutrients and solids from domestic waste
Industrial Waste	Toxic heavy metals along with hazardous organic and inorganic effluents
Agricultural Waste	Fertilizers, pesticides, insecticides, herbicides, leaches from agricultural land containing nitrates, phosphates and potash pollute the ground water.

Note: Ground water can also be polluted by septic tanks, refuse dumps.

Harmful Effects of Ground Water Pollution

Polluted groundwater is a major cause for the outbreak of epidemics and chronic diseases like typhoid, jaundice, dysentery, diarrhoea, tuberculosis, and hepatitis. Polluted ground water for irrigation damages the crops and decrease the yield of agricultural products.

In view of the above, it is extremely important to protect the ground water from getting polluted. This can be achieved by taking appropriate steps for predisposal treatment of toxic industrial effluents and municipal discharges. It is also important that the location of wells for drinking water supplies should be carefully decided.

Surface Water Pollution

Water present in rivers, streams and lakes constitute the surface water. It is present to the extent of only 0.0091 per cent of the total global water (see Table 2.8). All segments of our society (urban, rural, industrial, agricultural) are responsible for surface water pollution. Various materials and processes may pollute fresh water bodies. Some of these sources are given in Table 2.8.

Table 2.8 Some Sources of Surface Water Pollution

Sources of contamination	*Contaminants discharged*
Agricultural run-offs	Agrochemicals (pesticides herbicides, fungicides)
Accidental spills of chemicals	Different chemicals
Leaks from surface, storage tanks or	Gasoline, oil pipelines
Run off from industrial sites like factories, mines etc.	Solvents, chemicals
Radioactive material processing plants	Radioactive materials

The chemical composition of surface water (river water) is given in Table 2.3. The maximum permissible limits of effluents discharged in surface water is given in Table 2.9 below.

Table 2.9 Permissible Limit of Various Characteristics of Effluents

Characteristic of effluents	*Maximum permissible limit*
pH	5.9
BOD	20 mg/L
Suspended solids	30 mg/L
Particulates	15 mg/L
Cyanide	0.1 ppm
Phenols	0.5 ppm
Sulphate	1 ppm
Toxic metals	0.5 ppm

The harmful effects of surface water pollution are similar to that of ground water pollution. Of course, the effects depend on the extent of pollution. Attempts should be made so that the surface water is free from the following contaminats.

- Compounds imparting colour, odour, taste and turbidity, e.g. oils, grease, phenols, toxic materials.
- Substances which float on the surface as oil, scums, etc.
- Toxic radionuclides.
- Materials which are harmful to acquatic flora and fauna.
- Thermal effluents which should not be discharged in the surface water.
- Chlorinated compounds.

Water, as we know is the only natural resource available in plenty and is essential for survival of all living organisms. Fresh potable water is available only to the extent of about one per cent in rivers, lakes and as ground water. However, clean (or potable) water has become a precious commodity since its quality is degraded by various sources of pollution. Some of the important sources are

- Sewage and domestic wastes
- Industrial effluents
- Agricultural discharge/run off
- Soap and detergents

Sewage and Domestic Wastes

The word sewage is generally used for combined liquid wastes discharged from all domestic and industrial sources (toilets, etc.). It consists of mixture of all liquid domestic wastes, including faecal matter, urine, soap and detergents.

The total spent water (including sewage) of the community flowing through a sewer is referred to as **municipal waste**. There is uncontrolled dumping of wastes from rural areas, towns and cities into water bodies. Due to extensive population explosion, tremendous quantities of sewage are generated. Most of the sewage or municipal waste receives no treatment, particularly in the underdeveloped countries and some developed countries of the world. It is believed that 75 per cent of water pollution is caused by sewage and domestic wastes.

Characteristics of Sewage

The characteristics of sewage include both physical and chemical characteristics.

Physical Characteristics

- **Appearance.** The sewage water normally has a high turbidity and appears grayish. At times the colour becomes black due to partial decomposition.
- **Odour.** Sewage after some times develops a characteristic rotten egg odour of hydrogen sulphide. This is due to putrification.
- **Temperature.** Sewage is normally a few degrees higher than the water supply.

Chemical characteristics

The chemical characteristics of sewage can be of two types, vig. inorganic or organic.

- **Inorganic.** The common mineral matter comprises of nitrates, phosphates, sulphates, carbonates and chlorides of calcium, magnesium, sodium, potassium and iron. Besides, some trace quantities of some toxic metals like copper, chromium, zinc, manganese lead and nickel may also be present.
- **Organic.** It includes fatty acids, esters, amino acids, amides, amino sugars and their decomposition products. These constituents affact the pH and the odour.

The municipal wastes must be suitably treated before being discharged in water bodies. This aspect forms the subject matter of a subsequent section.

Industrial Effluents

Industrial waste water results from all manufacturing industries. These discharges are contaminated with materials used in the plant (starting materials), manufactured end products, intermediate products and by-products. All these materials are responsible for pollution of water bodies. The type of pollutants depend on the various industrial processes.

Agricultural Discharges

Most of the agrochemicals, viz., fertilizers, insecticides, herbicides and fungicides are used in large quantities for better and quality crops. These agrochemicals are made by various industrial manufacturing processes. The discharges from these industries contain various toxic chemicals and when discharged into water bodies are responsible for a number of adverse effects.

Fertilizers are essential plant nutrients. When fertilizers containing sulphates, nitrates and phosphates are added to the soil, some of these are washed off through rain fall, and these agricultural run off find their way into various water bodies. The excessive use of fertilizers disrupt the nitrogen and phosphate balance in the water bodies, resulting in increase of BOD and this may lead to **eutrophication** of lakes, etc.

Some pesticides, insecticides and herbicides are not biodegradable. When sprayed on crops, they remain adhered to the soil for long periods. During rainfall, these agrochemicals are carried as suspended particles into water bodies and create problem.

Soaps and Detergents

Soaps are sodium salts of fatty acids and are obtained by saponification of fats and oils. Synthetic detergents, on the other hand, are soap substitutes and comprise of surfactant, builder and other ingridients. Soaps are not effective in hard water, since there is considerable loss of soap before it is available for cleaning. However, detergents can be used in hard water. A common class of detergent is alkyl benzene sulfonates. Both soap and detergents contain hydrophobic (water repelling) or non-polar group (the fatty acid chain in soaps and the alkylbenzene group in detergents) as well as hydrophilic (water attracting) or polar group (the carboxylate group in soap and the sulphonate group in detergents). It is due to the presence of both hydrophobic and hydrophilic groups, that both soaps and detergents lower the surface tension of water and aid in wetting of clothes, dishes etc. and in emulsifying and dispersing of soil or dirt particles; such substances are called **surfactants**.

The soaps are biodegradable and so do not cause any environmental problems. In aquatic system, soaps are degraded by microorganisms in presence of oxygen to harmless products (CO_2, H_2O, carbonates and bicarbonates). The main disadvantage is that the soaps are not effective in hard water; this leads to wastage of soaps.

$$2R\,COO^{-}\,Na^{+} + Ca^{2+} \longrightarrow (R\,COO)_2\,Ca + 2Na^{+}$$

$$2R\,COO^{-}\,Na^{+} + Mg^{2+} \longrightarrow (R\,COO)_2\,Mg + 2Na^{+}$$

The detergents, on the contrary are not completely biodegradable. So some of the detergents always persist in water bodies and causes environmental problems. It is found that detergents having branced chains are less biodegradable than those having linear carbon chain. Some examples are:

Linear Chain Detergents (More Biodegradable)

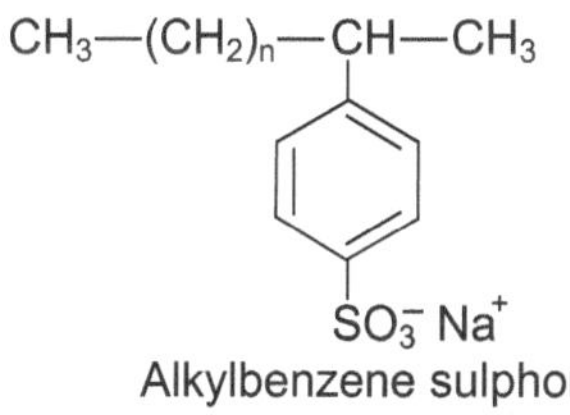

Alkylbenzene sulphonates

$$CH_3—(CH_2)_{12}—CH_2—OSO_3^{-}\,Na^{-}$$

Teepol

Branched Chain Detergents (Less Biodegradable)

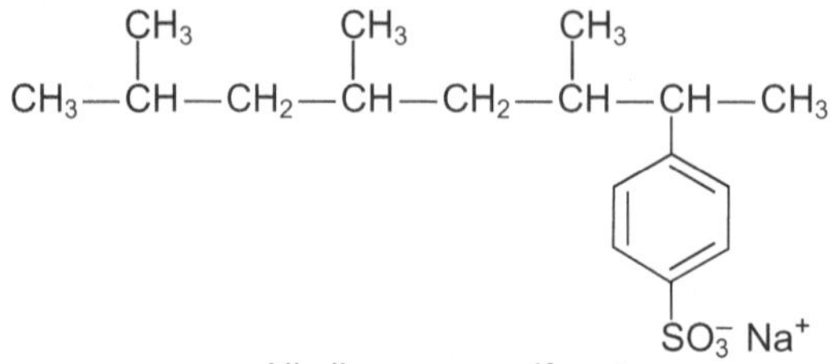

Alkylbenzene sulfonates

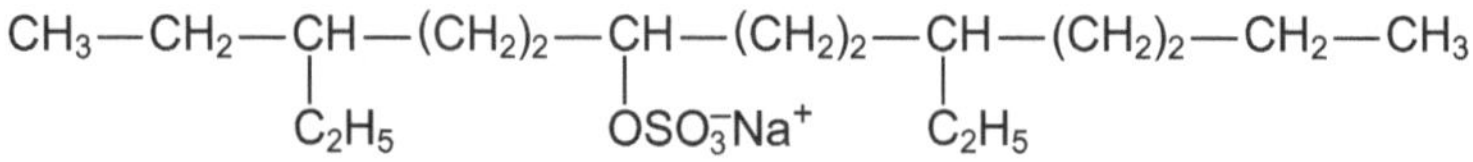

Tengitol–7

Types of Surfactants

The surfactants can be anionic, cationic, non-ionic or amphoteric in nature. Table 2.10 gives various types of surfactants along with their applications.

Table 2.10 Types of Surfactants

Types	*Examples*	*Uses*
Anionic	Soaps 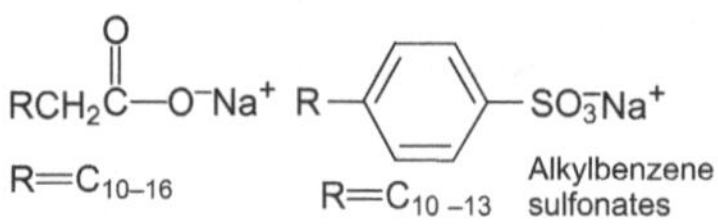	Most commonly used for toilet soap, laundry detergents and shampoos.
Cationic	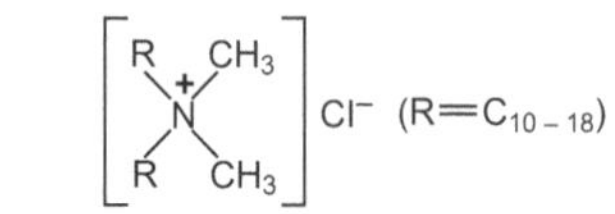 Quaternary ammonium salts 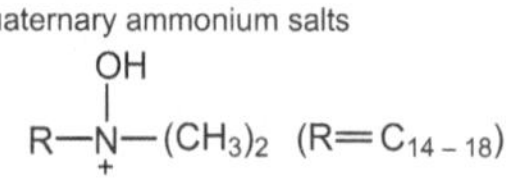	Used mostly as fabric softners
Non-ionic	$R(CH_2CH_2O)_n$ H Polyoxyethylenes ($R=C_{6-16}$; n=3–15)	Used for heavy duty laundry detergents
Amphoteric	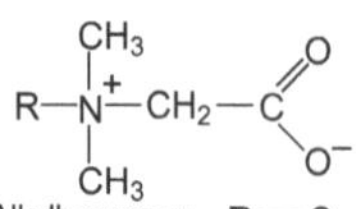 Alkylbenzenes $R=C_{12-18}$	Used for shampoos

Ingredients of Detergents

As already stated, the detergents are of synthetic origin and contain surfactants (mostly anionic like alkyl benzene sulphonates) and builder along with other additives. Some of the additives are given in Table 2.11.

Table 2.11 Additives of Detergents

Additive	***Example***	***Role of additives***
Builders*	Polyphosphates	Better cleaning action even in hard water
Brighteners	Ultramarine	Improve the appearance of the fabric
Bleachers	Perborates	Removing stains etc.
Enzymes	Amylase	Removing starch, protein based stains
Softeners	Quaternary ammonium salts	Making fabric feel soft

Note: Builders, the most important additive will be discussed in detail subsequently.

Builders are the most important constituents of detergents and are present to the extent of about 40 per cent of the total weight. Following are given the most useful and powerful builders.

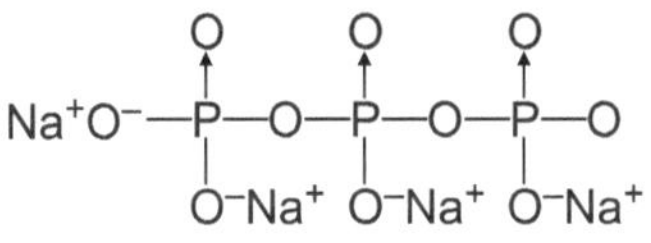

Sodium tripolyphosphate

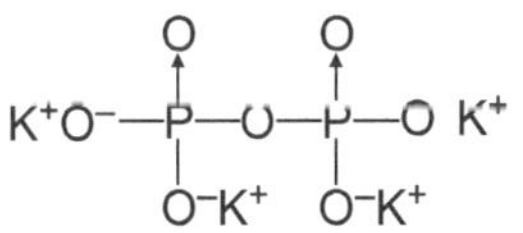

Potassium pyrophosphate

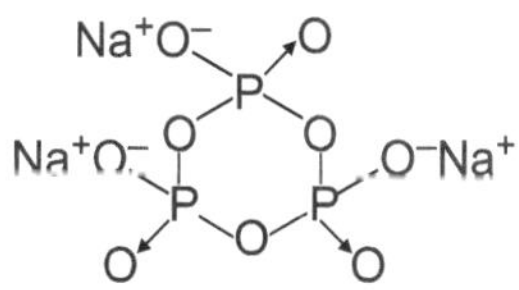

Sodium trimetaphosphate

Following are given some of the common and useful functions of builders.

i. The builders act as sequestering agents. These block the Ca^{2+} and Mg^{2+} ions present in hard water.

$$K_4P_2O_7 + 2\ Mg^{2+} \longrightarrow Mg_2(P_2O_7) + 4\ K^+$$

It is due to this reason that detergents are effective even in hard water.

ii. Builders along with surfactants exhibit marked synergistic effect; this makes their combination a much better cleaning agent.

iii. Builders make the wash solutions alkaline–a requisite necessary for effective removal of soil and dirt particles.

iv. Builders act as buffers. This implies that change of pH during washing is taken care of.

v. Builders are responsible to make the surfactant molecules in a concentrated wash solution to aggregate in groups, commonly called **micelles**. These micelles remove oil and grease particles (from the fabric to be cleaned) by dissolving them.

The main problem with polyphosphate additives (builders) is environmental. These cause eutrophication of water bodies. This aspect is discussed in a subsequent section.

Effects of Soaps and Detergents on Environment

Soaps, as already stated, are biodegradable. This implies that when discharged into water bodies, these are acted upon by microorganisms in presence of dissolved oxygen into harmless products (e.g. carbon dioxide, water, carbonates and bicarbonates). On the other hand, detergents affect the environment. Following are given some of the harmful effects of detergents on the environment.

i. Detergents containing phosphate as builders contribute to the growth of algae in water bodies. This results in consuming the dissolved oxygen and the aquatic animals are deprived of oxygen resulting in their death. Thus, phosphate detergents are the main cause of eutrophication of water bodies, particularly lakes.

ii. Aquatic animals and water plants are severely affected by synthetic detergents. A concentration of 2.5–3.5 per cent of common household detergents causes mortality to some types of fish and retard the growth of (or even kill) the water plants.

iii. Health hazards–The workers in detergent industry are exposed to enzyme additive in the formulation section. The enzymes attack the proteins in the soft lung tissue resulting in allergic conditions and respiratory problems.

iv. The potable water acquires objectionable taste.

v. The detergents make the water purification process difficult. This is because, the microorganisms which oxidize organic pollutants to harmless materials are killed. Also, with detergents, the oily and greasy impurities are emulsified and their removal becomes difficult by skimming process.

The best way to check the environmental problems is to use non-phosphate builders. Some builders which have been tried are EDTA, sodium citrate, zeolite-A [$Na_2O.Al_2O_3.2SiO_2.4.5H_2O$] and nitriotriacetic acid [$N(CH_2COOH)_3$]. However, none of these is as efficient as the phosphate additives.

Sea Water Pollution

Oceans are the major source of water in the world. The discharge of waste substances into the sea, which result in harming aquatic organism, human health and impairment of the quality of seawater is called **marine pollution**. The water bodies, *viz.*, rivers, seas and oceans are used by mankind as a dustbin for dumping all types of waste materials. These include industrial waste, heavy metals, agrochemicals, mine tailings, urban and rural sewage, farm and forest run offs. Even radioactive wastes are dumped into the seas. As in the case of river water pollution, the marine pollution is also caused by sewage and domestic wastes, industrial effluents, agricultural discharges and soap and detergents. These have already been discussed earlier. Besides these the marine pollution is caused by:

- Oil pollutants and oil spills
- Pollution by radioactive materials.

Oil Pollutants and Oil Spills

These are the major cause of marine pollution. At present, the annual consumption of refined petroleum products all over the world is more than 50 billion barrels (about 1600 billion gallons). This consumption is associated with some losses which can be accidential or intentional and are responsible for marine pollution. It is estimated that the total annual influx of petroleum oil into the oceans is more than 250 million gallons. The discharge of petroleum oil or its products into the oceans occurs at various stages of petroleum industry. These include :

- Extraction or isolation of crude oil from underground sources which can be below surface of earth or oceans.
- Transportation. The crude oil is transported through pipelines or oil tankers. Leakage may occur in both the modes of transportion. Transportation in huge oil tankers through the seas is the major cause of marine pollution. The oil tankers may spill oil in the oceans or may meet with an accident.
- Refining of petroleum oil in various oil refineries.
- Manufacture of petrochemicals which are made from various fractions of petroleum refining.
- The oil tankers are cleaned periodically with water and the washings discharged into the seas. About 3 million tons of oil is discharged annually by the process.

Some Major Oil Tanker Accidents Massive marine pollution in the sea is caused by oil tanker accidents. The National Academy of Sciences, Washington has estimated that about 8–9 million metric tons of oil enters the oceans annually. Some of the major oil tanker accidents in the seas are given below.

- A Liberian tanker, Torrey canyon got wrecked (1967) on the Ollard Rock near Land End, England resulting in the discharge of about 100,000 tonnes of oil into the seas.
- The oil tanker, M.V. Cosmos Pioneer broke into two (1973) near Porbunder in Gujarat, India due to rough weather and released about 18,000 tonnes of light disesel oil in the Gujarat coast.
- An American oil tanker, 'Transhuron' collided the atolls of the Kilton Islant in the Laccadives seas (1974) discharging about 5000 tons of furnace oil on the beaches.
- An oil tanker 'Urquiola' met will an accident (1976) near the Spanish coasts La Covuna and discharged about 100,000 tons of crude oil.

- The oil tanker 'Exxon Vaaldez' crashed on Prince William sound in Alaska (1989) discharging 11 million gallons of oil into Alaska's waters.
- During gulf war (1991) about 300 million gallons of oil in the Persian gulf was discharged in the water. The oil spill hit Saudi shores.

Effects of Oil Pollution

- The oily surface markedly reduce the rate of oxygen uptake by water thereby affecting aquatic life.
- Oily slit markedly decrease the penetration of sunlight. This retared the photosynthesis in aquatic region.
- Fishery is hampered.
- Soluble aromatics present in oil affect the aquatic organisms. Some of the aromatic compounds are carcinogenic and affect plant metabolism. The aromatic compounds are harmful to mankind. These produce irritation, anesthesia and damage liver and kidneys.
- Hundreds and thousands of birds, being vulnerable to damage of oil spills in the oceans die due to effects of oil.

Control of Oil Pollution in Seas

Some of the method to control the spreading of the oil spills are:

- *Skimming*. This is the simplest and best method. This method should be used in shortest possible time after the spilling has taken place, otherwise the wind will spread the oil spills over a very large surface making skimming difficult.
- *Use of Absorbents*. Some suitable absorbents like polyurethane, saw dust, chopped straw are spread on the surface of oil and then skimmed off.
- *Burning of oil slick*. This method, through useful causes extensive air pollution and may also cause thermal pollution.
- *Chemical Additives*. These are used to solidify oil from water surface and then the solid material is easily skimmed.
- *Microorganisms*. Certain oil eating microbes were used in cleaning up oil spills in the waters of the state of Texas, USA.

Pollution by Radioactive Materials

Radioactive pollutants enter the water bodies (especially seas and oceans) from various sources, some of which are :

- Mining and processing of ores to produce unstable radioactive substances
- Nuclear Power Plants
- Nuclear reactions
- Nuclear weapons
- Use of radioactive isotopes is medical, industrial and research applications.
- Testing of nuclear weapons in air, under water give radioactive fallout.

Following are given some of the bad effects due to radioactive pollutants in water.

- they cause serious skin cancer, carcinomas, melanomas, breast cancer, leukemia, breakage of DNA and catracts.
- they cause radiation sickness, which is characterized by nausea, vomiting, diarrhoea, epilation and lethargy.
- they cause genetic dissolvers, some gene mutations and blood abnormalities.

2.5.4 Impact of Aquatic Pollution on Human Health

We now know that water is easily polluted by physical and chemical agents present in the atmosphere, microbial disease causing germs, nuisance organisms like slime, molluses, sewage and industrial effluent and agricultural based effluents. All these pollutants effect our health by causing various diseases. They also effect our other resources such as animals, forests etc. on which we directly and indirectly depend for food and survival.

Water as a Carrier for the Transmission of Diseases

A number of microorganisms (e.g. bacteria, viruses, protozoans, insects and helminthes etc.) present in polluted water can cause infectious diseases. These diseases are called water-borne diseases. These water-borne diseases can be present in the form of outbreaks. This usually happens if the dose of contamination is heavy. Cholera and Jaundice (hepatitis) usually occur in the form of outbreaks, which can be very explosive. However, when contaminated water is treated or not used, the outbreak subsides (e.g. in the case of cholera). The following table gives diseases caused by microorganisms commonly in the polluted water. These diseases are an important cause of ill health particularly among the people of the developing countries.

Table 2.12 Diseases Caused by Microorganisms Commonly in Polluted Water

Organism	*Diseases*	*Remarks*
Ascaris sp.	Nematode worms	Danger to man from polluted water and dried sludge used as a fertilizer.
Bacillus anthracis	Anthrax	Found in waste water spores are resistant to treatment.
Brucella sp.	Brucellosis, Malta fever in man, contagious abortion in sheep, goat and cattle.	Transmitted by infected milk by contact, by waste water.
Entameoba histolytcca	Amoebic dysentery	Spread by contaminated water.
Leptospira sp.	Leptospirosis (Weil's Disease)	Caused by sewer rats.

(Table 2.12 Contd...)

Organism	*Diseases*	*Remarks*
Mycobact. tuberculi	Tuberclosis	Isolated from waste water
Salmonella paratyphi	Paratyphoid fever	Common in waste water
Salmonella typhi	Typhoid fever	Common in waste water
Salmonella sp.	Food poisoning	Common in waste water
Shigella sp.	Bacillary dysentery	Polluted water
Vibrio cholarae	Cholera	Polluted water
Virus	Poliomyelitus hepatitis	Waste water treatment plants
E. coli	Diarrhoea	Polluted water

Chemical Toxicants in Water Affecting Health

We are aware that there are a number of chemicals in drinking water. These may be naturally occurring or reach the water due to human activities. The presence of these chemicals in water do not pose health problems if their levels are below the threshold level. However, if these toxicants are present in excessive amounts, their presence is a matter of great concern since excessive amounts are detrimental to the health. Various chemicals which may pose a threat are: Fluoride, Arsenic, Lead, Nitrates, Pesticides, Cadmium, Chromium and Cobalt. Let us see the effects of these chemicals one by one.

Fluoride

The major source of fluoride to man is drinking water. In most parts of India the fluoride content is about 0.5 mg/L. There are, however, some areas where fluoride content in drinking water may be as high as 3 – 12 mg/L. The recommended level of fluoride in drinking water accepted in India in 0.5 to 0.8 mg/L.

High intake of fluoride causes humped back. Continuous consumption causes stiffness of bone joints, particularly spinal cord. It accumulates in bones causing outward bending of legs from knee-called **Knock Knee Syndrome**. Water rich in fluoride cause endemic fluorosis. It was first detected among cattles by the farmers of Andhra Pradesh in early 1930's.

Excess fluoride is removed by Nalgond techniques. It involves the addition of lime and alum in sequence followed by flocculation, sedimentation and filtration.

Arsenic

Arsenic is introduced into water via the dissolution of minerals and ores, from industrial effluent and from atmospheric deposition. A provisional guideline value for arsenic in drinking water of 0.01 mg/L. World's worst affected areas due to Arsenic contamination is the Indo-Bangladesh region. WHO has alarmed Govt. of India to keep Arsenic related disease under immediate national surveillance, giving it a status similar to that of cancer. WHO prescribed limit to 0.01 mg/L.

At higher levels, As(III) compounds coagulates proteins, possibly by attacking S-bond retaining the secondary and tertiary structure of proteins. In the biological systems, Arsenic compounds complexes will co-enzymes resulting in coagulation of biological proteins and uncoupling at phosphorylated sites. The effect on human depends on the arsenic dosage. Thus

Mild Dose – nausea, fainting, salivation, vomiting, burning pain in stomach.

Higher Dose – diarrhoea, peripheral neuritis, hyperketosis, conjuctivitis.

Chronic Dose – severe gastroenteritis, loss of weight, skin lesions, black food disease.

Lethal dose – death due to shock and vascular failure.

Arsenic is a cumulative poison. In animals, it affects the organs like liver and lungs causing toxicity to biological systems. Arsenic in water may be carcinogenic in nature.

Lead

The important sources of lead for biological systems are mining, smelting, winning of metals, automobiles, etc. In urban areas, citizens are exposed to it through emission of vehicles and food. About 200–300 mg/day of lead intake through food and about 10–15 mg/day through air and water is quite common in urban settlements. Lead is present in tap water to some extent as a result of dissolution from natural sources but mainly from household plumbing systems containing lead pipes. It is satisfying to note that emissional lead from vehicular emission has ceased since leaded petrol is no more used in cities like Delhi.

When present in < 40 mg/100 mL of blood, lead is considered as normal. However >100 mg/100 mL of blood is unacceptable.

Lead gets deposited in bones and is considered as a cumulative poison. Lead is a cumulative poison. It enters the body through inhalations, ingestion and absorption through the skin. On an average natural water contains about 20ppb of lead. It enters water bodies from mining, smelter discharges lead plumbing and weathering of rocks. After entering the human body, it gets concentrated in the bones, from where it enters the blood stream.

Lead interferes with hence synthesis leading to hematological damage. It interferes with the production of red blood cells and can also damage the brain, kidney and liver. Lead is particulars affects children and pregnant women; in the later case it may be responsible for miscarriage and birth defects.

Nitrates

Nitrates and Nitrites both are formed during the natural process of fixation of atmospheric nitrogen. These also enter the water bodies from agricultural farms where these are used as fertilizers. Normally these are present in natural waters to a small extent, but their presence in excessive amounts in drinking water is extremely harmful.

The danger of drinking water containing high nitrate content has long been recognized. Live stocks have been effected by well water containing 75–100 ppm of nitrate nitrogen. Human infants have been especially susceptible, bottle feed infants, of less than 3 months of age are most susceptible.

Nitrate poisoning brings about a condition called **methaemoglobinacemia**, resulting from change in the haemoglobin of the red blood cells that reduce their capacity to carry oxygen. This is referred to as blue baby syndrome.

The Table 2.13 given below gives the toxic and pathological effects of some heavy metals.

Table 2.13 Hazardous Effects of Some Metals on Human Beings

Metals	*Impact*
Arsenic	– Adversely affect peripheral circulatory systems and CNS, hyperkeratosis, ulcer in gastrointestinal tract, liver cirrhosis.
Barium	– Abdominal pain, diarrhoea, vomiting, excess salivation.
Cadmium	– Diarrhoea, growth retardation, malfunctioning of urinogenital system, CNS, Kidney, Liver.
Copper	– Sporadic fever, uremia, coma, hypertension.
Cobalt	– Diarrhoea, lowering blood pressure, paralysis, respiratory problems, deforming skeleton.
Chromium	– Generally hexavalent form, malfunctioning of gastrointestinal, CNS and urinogenital system.
Mercury	– Haemolysis, diarrhoea, severe headache, paralysis, desiccation of intenstinal epithelium, ultimately death, loss of appetite, convulsions, brain damage, malfunctioning of liver, kidney, anaemia, vomiting.
Zinc	– Abdominal cramps, vomiting, diarrhoea, renal damage.

Water as a Vector Habitat in Disease Transmission in Humans

Following are given the most important water-related vector borne diseases

- Malaria
- Schistosomiasis (found mostly in sub-Sahara Africa)
- Dengue fever
- Filariasis
- Japanese encephalitis

2.6 EUTROPHICATION

When sewage and agricultural run-off containing phosphates (phosphates are major ingredients of most of the detergents) or other nutrients enter natural water bodies, they cause over nutrition leading to eutrophication. In such a situation the water bodies became highly productive or eutrophic and the phenomenon is called eutrophication. The ponds or lakes, etc., during their early stages of formation are comparatively barren or nutrients-deficient, thus, supporting no or poor aquatic life. This stage of these water bodies is known as Oligotrophic. With the introduction of nutrients, there is a stimulated luxuriant growth of algae in water. This growth depletes the dissolved oxygen and affects the other aquatic flora and fauna and ultimately the lake dies.

2.6.1 Water Quality Parameters

No single parameter is possible to determine the quality of water. Depending on the use for which water is intended to be used, it should have the right parameter. For example, if water is to be used for washing purposes (laundry etc.), it should be ascertaind that it is not hard. Hard water is also harmful for use in boilers etc., as it leads to scale formation. For drinking purposes, water must be pure and free of any pollutants. Tables 2.14 and 2.15 give standards for industrial and drinking (human consumption by BIS).

Table 2.14 BIS Standards for Water for Industrial and Drinking Purposes

Parameter	***BIS standard***
Colour	Colourless
Odour	Odourless
Taste	Light, sour-sweet
pH	6–8.5
Specific conductance	300 μ mho cm^{-1}
Dissolved oxygen	4–6 ppm.

Table 2.15 BIS and USPH (United States Public Health Drinking Water) Standard for Potable Water

Contaminant	***Maximum contaminated level (ppm)***	
	BIS Standard	***USPH Standard***
Chloride	600	250
Sulphate	1000	250
Cyanide	0.0001	0.05
Fluoride	3	1.5
Nitrate + Nitrite	12	< 10
Phosphate	0.1	0.1
Calcium	100	100
Magnesium	30	30
Barium	1	1.0

(Table 2.15 Contd...)

Contaminant	Maximum contaminated level (ppm)	
	BIS Standard	USPH Standard
Copper	1	1.0
Arsenic	0.002	0.05
Lead	0.1	< 0.05
Iron (filterable)	0.3	< 0.3
Chromium	0.05	0.05
Zinc	0.05	5.5
Pesticides (total)	0.005	0.005
Total bacterial count	1 × 106	1 × 106

The quality of water is basically judged by three parameters, viz. Dissolved Oxygen (DO), Biochemical Oxygen Demand (BOD) and Chemical Oxygen Demand (COD). However, some other parameters (given below) are also useful for determining the quality of water:

Alkalinity

Most Probable Number (MPN)

Total Solids

Oxidation

Transparency

Silica content

Hardness

Metals

Content of chloride, sulphate, phosphate, nitrate and microorganisms.

(*i*) Dissolved Oxygen (DO)

It is the amount of oxygen gas that is dissolved into water at any source. The degree of self purification is high if the amount of DO is high. However, if the amount of DO is less, all the impurities in water will not be eradicated by the self-purification process; in such cases better purification procedures are necessary. In fact, the value of DO in water is a useful measure of the condition of water (or the degree of pollution).

In water, oxygen may be directly absorbed from the atmosphere. It may also be released in water by aquatic plants during respiration or photosynthesis by aquatic plants. The amount of oxygen that can be dissolved in water decreases as the content of soluble salt increases. Due to this, the availability of oxygen is comparatively less in sea water than in river water. Also, the lower the temperature, the more will be the amount of DO. The solubility of oxygen in water is also directly proportional to the atmospheric pressure. Thus, a lake or pond at a higher level will have a lower value of dissolved oxygen than the one near the sea level. Table 2.16 gives the level of oxygen or the amount of oxygen that can be dissolved in water at different temperatures.

Table 2.16 Dissolved Oxygen in ppm (Saturation Value) in Fresh and Sea Water

	Temperature °C					
	0	5	10	15	20	25
Fresh water	14.6	12.8	11.3	10.2	9.2	8.4
Sea water	11.3	10.0	9.0	8.1	7.4	6.7

The shortfall in DO or the degree of unsaturation is determined by substracting measured value of DO at a given temperature from the saturation value of DO at the same temperature by referring to the above table. The DO valuc in any sample of water is determined as follows.

Determination of Dissolved Oxygen

The dissolved oxygen is allowed to react with potassium iodide. The liberated iodine is then titrated with standard $Na_2S_2O_3$ solution. Since the dissolved oxygen in water remains in molecular state, it is not capable of undergoing reaction with KI. Therefore an oxygen carrier, e.g. manganese hydroxide is used to bring about the reaction. The manganese hydroxide is produced in situ by the action of potassium hydroxide and manganous sulphate. The reactions involved are:

$$MnSO_4 + 2KOH \longrightarrow Mn(OH)_2 + K_2SO_4$$

$$2Mn(OH)_2 + O_2 \longrightarrow \underset{\text{basic manganic oxide}}{2\,MnO\,(OH)_2}$$

$$MnO(OH)_2 + H_2SO_4 \longrightarrow MnSO_4 + H_2O + [O]$$

$$2KI + H_2SO_4 + [O] \longrightarrow K_2SO_4 + H_2O + I_2$$

$$I_2 + 2NaS_2O_3 \longrightarrow 2NaI + \underset{\text{Sodium tetrathionate}}{Na_2S_4O_6}$$

It is found that 5 mL of 0.025 M $Na_2S_2O_3$ correspond to 1 mgL^{-1} DO.

Interferences due to oxidizing agents such as NO_2^- and SO_3^{2-}, present in waste water may be eliminated by addition of sodium azide to the alkaline solution. On acidification NO_2 is decomposed.

$$2NaN_3 + H_2SO_4 \longrightarrow 2\,HN_3 + Na_2\,SO_4$$

$$HNO_2 + \underset{\substack{\text{Hydrazoic}\\\text{acid}}}{HN_3} \longrightarrow N_2O + N_2 + H_2O$$

(*ii*) Biochemical Oxygen Demand (BOD)

Biochemical oxygen demand (BOD) is a measure of oxygen that would be needed by the microorganisms to dicompose the organic and inorganic pollutants in polluted water. The microorganisms transform the pollutants into non-hazardous products (CO_2 and H_2O). During this process dissolved oxygen is consumed.

$$\underset{\text{Organic matter}}{CH_2O} + O_2 \xrightarrow[\text{Aerobic oxidation}]{\text{Microorganisms}} CO_2 + H_2O + \text{Bacterial cell}$$

The inorganic pollutants may include sulphites, sulphides, thiosulphates and ferrous iron. These pollutants are attacked by oxygen in presence of enzymes released by microorganisms.

Determination of Biochemical Oxygen Demand

For determination of BOD, the water sample is first saturated with oxygen and then incubated at constant temperature (usually 20°C) for five days. During this time the microorganisms in polluted water oxidise the pollutants completely. After five days, the remaining amount of dissolved oxygen is determined and the BOD is obtained by substraction.

$$\text{BOD} = \begin{bmatrix}\text{Saturation value of dissolved} \\ \text{oxygen per litre at 20°C}\end{bmatrix} - \begin{bmatrix}\text{Amount of } O_2 \text{ dissolved} \\ \text{per litre at 20°C after 5 days}\end{bmatrix}$$

The result is called the 5-day BOD, and is expressed in milligram of oxygen per litre of water (mg/litre) or in ppm. The 5-day BOD analyses is considered as an accepted standard test. The BOD of 75 mg/litre of polluted water indicates that the biodegradation of organic matter in one litre of a sample consumes 75 mg of oxygen. Drinking water should have a 5-day BOD of less than 1.5 mg/litre. The BOD of raw sewage ranges from 200 – 400 ppm, indicating a strong sewage. The major contributors to BOD are the chemical industries, pulp and paper industries and the food processing industry. The BOD water status is given in Table 2.17.

Table 2.17 BOD – Water Status

BOD	***Status***
1 mg/litre	very clean water
2 mg/litre	clean water
3 mg/litre	fairly clean water
5 mg/litre	doubtful, not used for drinking and pharmaceutical preparations
10 mg/litre	contaminated water
15 mg/litre	not used for drainage and fish reproduction population
150 mg/litre	sewage discharge from houses
200 mg/litre	industrial waste water
350 mg/litre	waste water from paper industries
1000 mg/litre	food processing industries waste water
2000 mg/litre	dairy discharge waste water
10000 mg/litre	tannery effluent

The BOD is not a true indication of pollution, but gives an idea of the problems arising due to waste materials in the water supplies. BOD is directly proportional to the organic matter oxidised.

(iii) Chemical Oxygen Demand (COD)

Like BOD, chemical oxygen demand (COD) is also a measure of the amount of oxygen required to degrade or breakdown the organic matter. In BOD, the degradation of organic matter is effected by microorganisms but in COD, the degradation is effected by some chemical oxidizing agent like potassium dichromate. It is believed that biochemical oxygen demand (BOD) does not give a correct assessment of the pollution load in water. This is because some compounds like detergents are resistant to microbial degradation and some other pollutants like cellulose are slow to get oxidized. Also the toxic compounds, if present in polluted water may poison the microorganisms before they can act on pollutants. Lastly, the BOD determination takes too long time (5 days).

The above problem can be overcome by another analytical method called chemical oxygen demand (COD). In this case, the water sample is treated with an oxidising agent (usually potassium dichromate in acidic medium), which oxidises most of the polluting substances including those which are resistant to microbial oxidation. The unused potassium dichromate is determined by back titration with a suitable reducing agent like Mohr's salt. The amount of potassium dichromate consumed is determined by substraction.

$$\begin{array}{c}\text{Amount of}\\ K_2Cr_2O_7\\ \text{Consumed}\end{array} = \begin{array}{c}\text{Amount of } K_2Cr_2O_7\\ \text{initially added}\end{array} - \begin{array}{c}\text{Amount of } K_2Cr_2O_7 \text{ determined}\\ \text{by back titration}\end{array}$$

The amount of oxygen used in the oxidation can be calculated from the strength of $K_2Cr_2O_7$ consumed using the following equation:

$$K_2Cr_2O_7 + 4\,H_2SO_4 \longrightarrow K_2SO_4 + Cr_2(SO_4)_3 + 4\,H_2O + 3(O)$$

The results are expressed in terms of the amount of oxygen in ppm that will be required to oxidize the contaminants. On the basis of COD status, the quality of water can be found out (see Table 2.18 below).

Table 2.18 COD Status

COD in mg L^{-1}	*Status*
0–5	Very clear water, used for drinking
5–20	Fairly clean
20–100	Unfit for drinking but can be used for washing and agriculture

COD determination is reliable and fast for assessment of the quality of water. However, it suffers from the drawback that aromatic hydrocarbon derivatives are not easily oxidized by most chemical oxidizing agents. So, their presence cannot be judged by COD determination.

(*iv*) Alkalinity

The total amount of substances that cause increased concentration of OH^- ions either on dissociation or as a result of hydrolysis is called **alkalinity** of water. In natural water, alkalinity is due to the presence of HCO_3^-, SiO_3^{2-}, $HSIO_3^-$ and CO_3^{2-}. Sometimes, the presence of salts of weak organic acids bind H^+ and thus result in increasing the concentration of OH^- ions rendering the natural waters alkaline.

The alkalinity may be regarded as **bicarbonate alkalinity** (A_b), **carbonate alkalinity** (A_c) or **hydrade alkalinity** (A_h), depending upon the presence of HCO_3^-, CO_3^{2-} and OH^- ions, respectively. Carbonate hardness is also responsible for the alkalinity of water. Alkalinity of water is determined by titrating it with a mineral acid using methyl orange as an indicator.

(*v*) Most Probable Number (MPN)

High population of bacteria like *E. coli* and *coliforms* is found in water polluted with organic wastes. Both *E. coli* and coliforms can be detected and enumerated with the help of MPN method. Coliform is present in human intestine and generally is not harmful but its presence indicates the presence of human waste in water. Polluted water shows high values of MPN.

(*vi*) Total Solids

The total solid content is the amount of non-volatile substances present in water in a colloidal and molecular dispersed state. It is expressed in milligram per kilograms (mg/kg). For determining the content of total solids in water, calcium and magnesium bicarbonates are converted into carbonates, viz. $CaCO_3$ and $MgCO_3$. In addition to the total solids, there are three other types of solids; these are:

i. Fixed residue solid

ii. Mineral residue solid

iii. Sulphate solids

Fixed residue solid is obtained by calcining the total solids for 10 – 15 min at 800°C; this results in removal of organic substances due to combustion and decomposition of carbonates. The residue obtained is the fixed residue solid content.

The quantity obtained by adding all cations and anions, including CO_3^{2-} and Al_2O_3, Fe_2O_3 and SiO_2 in water, is called the **mineral residue**. If the total solids are treated with concentrated sulphuric acid, all the cations present in water are converted into sulphates, the total mass of which constitutes the **sulphate content**.

(*vii*) Oxidation State

The contamination of water by organic substances is represented by oxidability and is expressed in milligram of oxygen required to oxidize the organic compounds contained in 1 kg of water. The oxidability is denoted as mg/kg. Oxidability can also be expressed by the amount of $KMnO_4$, mg/kg used to oxidize the organic substances. Oxidability, however, does not represent the total content of organic substances in water, since complete oxidation of all organic substances does not occur under the conditions in which oxidability is determined.

(*viii*) Transparency

It indicates the concentration of suspended solids, which can be determined by the weight method. The transparency can also be determined by the type **method** (cm) of a column of water in a glass tube, through which it is still possible to discern printing or by the **cross method** which two crossed black lines 1 mm thick are visible on a white paper placed on the bottom of a glass tube.

(*ix*) Silica Content

It is the concentration of silicic acid, H_2SiO_3, expressed as silica, SiO_2 contained in water. The concentration of SiO_2 in natural water varies from 5 – 10 to 90 mg L^{-1}. It is an indication of the extent of mineralization of water. Raw water contains silicic acid in the ionic ($HSiO_3^-$) as well as in colloidal state. The presence of silica in feed water for boiler units causes a number of operating conditions due to formation of silicate scales of low thermal conductivity; even heavy deposits on the blades and nozzles of turbines are formed. In modern practice, silica is removed from raw water using ion exchangers.

(*x*) Hardness

Water from all sources contains a great degree of dissolved mineral salts. Normally water contains dissolved magnesium and calcium bicarbonates, sulphates and chlorides in considerable amounts, along with traces of different cations and anions which introduce hardness to water. Water that produces or forms an insoluble precipitate when it is boiled or when soap is added is called hard water. In fact, **hard water** does not give lather with soap, such water may not always be unfit or harmful for drinking but cannot be used in laundry work and for steam generators in boilers. Water required for pharmaceuticals preparations, dairies, laundries and for boiler feed must be **soft water**. Water which gives lather easily with soap is called soft water. Before water is used, one must ensure the nature and amount of dissolved quantities present.

Soap reacts with soluble calcium or magnesium salts and forms an insoluble white scum or precipitate with hard water; this causes a loss of soap and reduces the cleaning action. A large amount of soap is needed to produce lather with hard water.

$$\underset{\substack{\text{Sodium stearate}\\\text{(soap)}}}{2\,C_{17}H_{35}COO^-Na^+} + \underset{\text{Hard water}}{CaCl_2} \longrightarrow \underset{\substack{\text{Calcium stearate}\\\text{(insoluble scum)}}}{(C_{17}H_{35}COO)_2Ca\downarrow} + 2\,NaCl$$

$$\underset{\substack{\text{Sodium alkanesulphonate}\\\text{(detergent)}}}{C_{15}H_{31}SO_3^-Na^+} + \underset{\text{Hard water}}{CaCl_2} \longrightarrow \underset{\text{Calcium alkane sulphonate}}{(C_{15}H_{31}SO_3)_2Ca\downarrow} + 2\,NaCl$$

Soft water does not contain dissolved calcium and magnesium salts. The presence of sodium compounds does not contribute to hardness in water. Depending on the nature of the anions present in it hardness of water is of following two types.

i. Temporary or carbonate hardness

ii. Permanent or non-carbonate hardness

Temporary or Carbonate Hardness

When water contains calcium or magnesium bicarbonate, the hardness is called **temporary hardness** or **carbonate hardness**. Such hardness can be removed by boiling, when calcium and magnesium bicarbonates from insoluble carbonates.

$$\underset{\text{Soluble}}{Ca(HCO_3)_2} \xrightarrow{\Delta} \underset{\text{Insoluble}}{CaCO_3\downarrow} + H_2O + CO_2\uparrow$$

$$Mg(HCO_3)_2 \xrightarrow{\Delta} \underset{\text{Insoluble}}{MgCO_3\downarrow} + H_2O + CO_2\uparrow$$

Temporary harness can also be removed by adding calculated quantities of lime. This method is called **Clark's method**. However, if too much lime is added, the water becomes again hard.

$$Ca(HCO_3)_2 + Ca(OH)_2 \longrightarrow 2\,CaCO_3\downarrow + 2\,H_2O$$

$$Mg(HCO_3)_2 + Ca(OH)_2 \longrightarrow MgCO_3 + CaCO_3 + 2\,H_2O$$

Permanent or Non-carbonate Hardness

It is due to the presence of soluble salts (chlorides and sulphates) of calcium, magnesium, iron or other heavy metals. Unlike temporary hardness, it cannot be removed by boiling. Carbonate hardness (temporary hardness) and non-carbonate hardness (permanent hardness) together constitute the **total hardness**. Before removing permanent hardness, the water should be boiled to ensure removal of temporary hardness.

2.7 DISSOLVED INORGANIC IMPURITIES

As the name suggests, these substances are dissolved in water. The nature of water depends on these dissolved inorganic impurities. Following table gives the dissolved impurities that may be present in water and their effects.

Table 2.19 Dissolved Inorganic Impurities

Nature of cations	*Impurity anion*	*Effects*
Calcium and magnesium	HCO_3^- CO_3^{2-}	Causes alkalinity and hardness
	Cl^- SO_4^{2-}	Causes hardness
Sodium	HCO_3^-	Causes alkalinity
	CO_3^{2-}	Causes alkalinity
	Cl^-	Causes change in taste
	F^-	Affects tooth enamel
Metal and their ions	Fe	Causes hardness, imparts red colour, change taste

(Table 2.19.Contd...)

Nature of cations	*Impurity anion*	*Effects*
	Mn	Imparts brown or black colour
	Pb	Causes lead poisoning
	As	Causes arsenic poisoning
Gases	O_2	Corrodes metals
	CO_2	Causes acidity, corrodes metals
	H_2S	Bad smell; acidity, corrosion

2.8 TOXIC METALS IN WATER

Natural waters and waste waters contain a number of trace elements, which are toxic at higher levels. Following table gives some of such elements along with their sources and effects.

Table 2.20 Some Toxic Elements in Water

Elements	*Sources*	*Effects and significance*
Arsenic	Mining by-product, pesticides, chemical waste	Toxic, possibly carcinogenic.
Cadmium	Industrial discharge, mining waste, metal plating, water pipes	Replaces zinc biochemically, causes high blood pressure, kidney damage, destruction of testicular tissue and red blood cells, toxicity to aquatic biota.
Beryllium	Coal, nuclear power and space industries	Acute and chronic toxicity, possibly carcinogenic.
Boron	Coal, detergent formulations, industrial wastes	Toxic to some plants.
Chromium	Metal plating, cooling tower water additive, (chromate) normally found as Cr (VI) in polluted water	Essential trace elements; possibly carcinogenic as Cr (VI).
Copper	Metal plating, industrial and domestic waste, mining, mineral leaching.	Essential trace elements, not very toxic to animals, toxic to plants and algae at moderate levels.
Fluorine (Fluoride ion)	Natural geological sources, industrial waste, water additive.	Prevents tooth decay at about 1 mg/L, causes mottled teeth and bone damage at about 5 mg/L.
Lead	Industry, mining, plumbing, coal, gasoline	Toxic (anaemia, kidney disease, nervous disorder), wild-life destroyed.

(Table 2.20 Contd...)

Elements	*Sources*	*Effects and significance*
Manganese	Mining industrial waste, acid mine drainage, microbial action on Manganese mineral at low pH	Relatively non-toxic to animals, toxic to plants at higher level, stains materials (bathroom fixtures and clothing).
Mercury	Industrial waste, mining, natural sources	Toxic to animals.
Selenium	Natural geological sources, sulphur, coal	Essential at low levels but toxic at higher levels
Zinc	Industrial waste, metal plating, plumbing	Essential in may metallo-enzymes, toxic to plants at higher levels

Besides the above, the water bodies contain a large number of pesticides which are discharged into water bodies by run offs from agricultural land.

Arsenic (III) compounds in higher concentrations coagulate proteins by attacking the sulphur bonds. The antidotes for arsenic poisoning are chemicals having –SH group, which binds to As(III). An example is 2,3-dimercaptopropanol.

$$\underset{\displaystyle \text{SH}}{\text{HSCH}_2\text{—}\underset{|}{\text{CH}}\text{—CH}_2\text{OH}}$$

2.8.1 Inhibition of Enzyme Action

The toxic chemicals (viz., metal ions) inhibit enzyme. Heavy metal ions, particularly Hg^{2+}, Pb^{2+} and Cd^{2+} attack the active sites (e.g. SCH_3 and –SH in methionine and cysteine amino acids, which are part of enzyme structure). Thus,

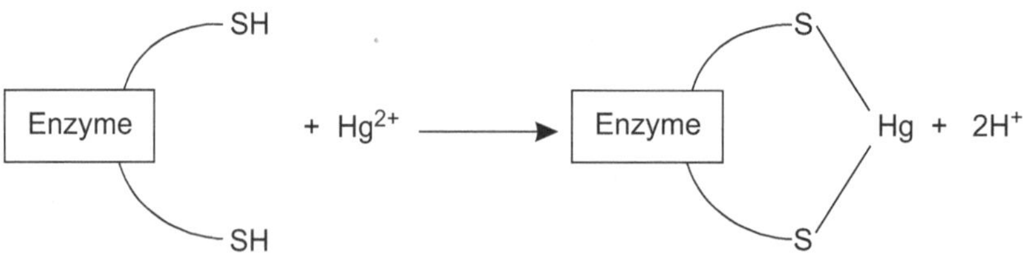

In this way, the enzyme action is inhibited. Also, in metalloenzymes, the metal of the metalloenzyme is replaced by another metal ion of similar size and charge. For example, Zn^{2+} in some metalloenzyme is replaced or substituted by Cd^{2+} leading to cadium toxicity. The enzymes, which are inhibited by Cd^{2+} include adenosine triphosphate, alcohol dehydrogenase, amylase, carbonic anhydrase. Pb^{2+} Inhibits acetylcholinesterase, alkaline phosphate, adenosine triphosphate, and carbonic anhydrase.

The biochemical effects of some toxic elements like arsenic, lead, cadmium and mercury from the subject matter of subsequent discussions.

(*i*) Arsenic

Arsenic occurs in insecticides, fungicides and herbicides. It finds its way into agricultural land and enters the food chain. Arsenic exerts its toxic action by attacking –SH groups of an enzyme resulting in inhibition of enzyme action.

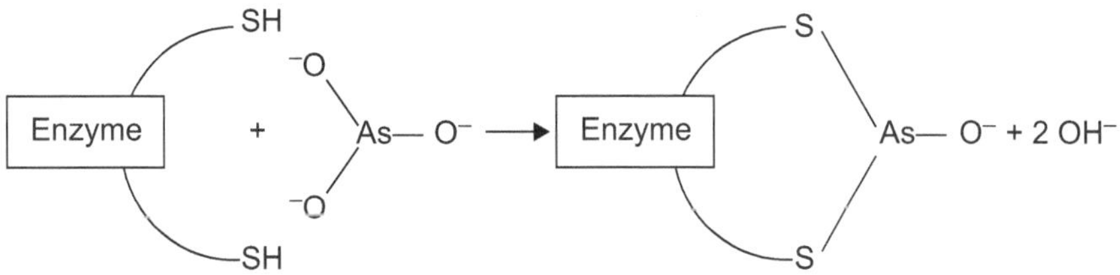

The enzyme, pyruvate dehydrogenase, which is responsible for generating cellular energy is deactivated by complexation with As(III). In this way, the generation of ATP is prevented.

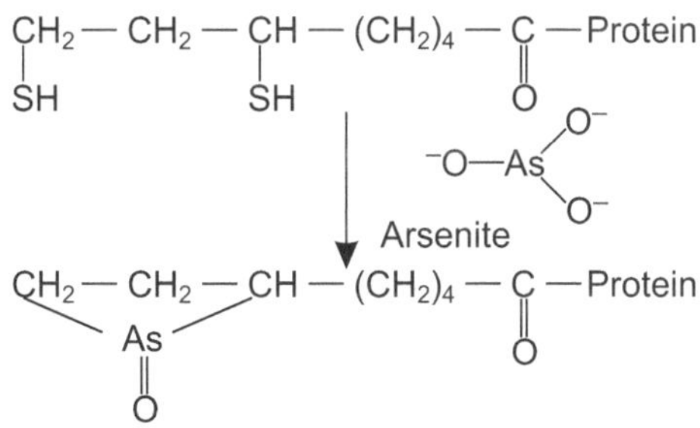

Arsenic also interferes with some biochemical processes involving phosphorus; this is due to chemical similarity of arsenic and phosphorus.

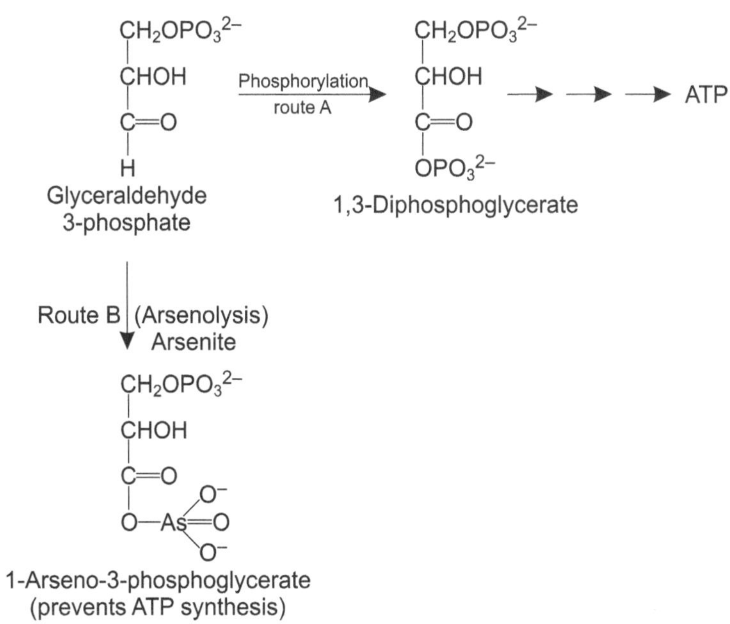

Thus, ATP (adenosine triphosphate), as we know, is generated from glyceraldehyde 3-phosphate by phosphorylation (route A). However, in presence of arsenic, glyceraldehyde 3-phosphate gives 1-arseno-3-phosphoglycerate (route B). Thus, in the later route (B), in place of phosphorylation, arsenolysis occurs, the product formed is 1-arseno-3-phosphoglycerate, which prevents ATP synthesis.

Arsenic occurs in water as a result of mineral dissolution, industrial wastes or the application of insecticides. The presence of as little as 100 mg in water may cause acute arsenic poisoning, because the usual level in potable water is 10 to 100 ppb. chronic effects can appear from the accumulation of arsenic in the body at low intake levels for prolonged periods.

Analytical Procedure for Estimation of Arsenic

The water sample is concentrated and arsenic is reduced to arsine (AsH_3) by zinc in acid solution in a gutzeit generator. The arsine is then passed through a scrubber containing glass wool impregnated with lead acetate solution and then into an absorber containing silver diethyldithiocarbamate [$AgSCSN(C_2H_5)_2$] dissolved in pyridine. A soluble red complex is formed as a result of the reaction of arsenic with the Ag salt of the complex, and this complex can be measured at 535nm. The presence of 0.002 to 0.1 ppm of arsenic can be accurately measured by this method.

(*ii*) Cadmium

Cadmium occurs in nature along with zinc in zinc blende (ZnS). Cadmium poisoning was first observed in Japan in the form of *itai itai* or 'Ouch Ouch' disease; in this disease, the bones become fragile. Higher levels of cadmium cause kidney problems, anaemia and disorders of bone marrow, hypertension, anaemia, renal dysfunction and even cancer. Most of the cadmium ingested into the body is trapped in kidneys and eliminated. However, a small fraction is bound effectively by the body proteins, metallothionein, present in kidneys. When excessive amounts of Cd^{2+} are ingested, it replaces Zn^{2+} at major enzymatic sites and thus, causes metabolic disorders.

The presence of cadmium in water sample may be detected by *p*-nitrobenzenediazoamino benzene (cadion) or by dinitrodiphenyl carbazide. In this procedure cadion (0.02g) is dissolved in alcoholic KOH solution (100ml of 0.002 N solution). A drop of this solution is placed upon a strip of filter paper resting on a thick piece of absorbent paper (blotting paper). A drop of the sample solution, rendered slightly acetic with acetic acid is added followed by addition at a drop of 10% KOH solution. A bright pink colour surrounded by a violet ring indicates cadmium. If Cr, Co, Cu, Fe, Mg or Ni ions are present, a few drops of saturated solution of Rochelle salt (sodium potassium tartaratie) is added. Silver, if present may be *precipitated* by adding 1 per cent KI solution and then filtered. If Hg is present, it may be precipitated with H_2S and then filtered. Cyanides are boiled off from the acid solution of the sample.

Alternatively, a drop of acid, neutral or ammoniacal solution of the sample is placed on a spot plate and a drop each of 8 per cent NaOH solution and 10 per cent NaCN solution are added. Finally a drop of 0.1 per cent alcoholic solution of the reagent (diphenyl carbazide) and 2 drops of 40 per cent HCHO are added. A blue green colouration or precipitate indicates the presence of cadmium. The reagent is red in alkaline solution, but turns violet in presence of HCHO. The limit of sensitivity is 0.0001 mg of Cd in a concentration of 20ppm.

Analytical Procedure for Estimation of Cadmium

Cadmium reacts with dithizone to form an intense pink to red coloured complex. The complex can be extracted in chloroform and then the extract is measured spectroscopically at an absorbance of 518 mm. The extraction is carried out from a strongly basic solution, Ag, Hg, Co, Cu etc interfere. These can be removed by preliminary extractions.

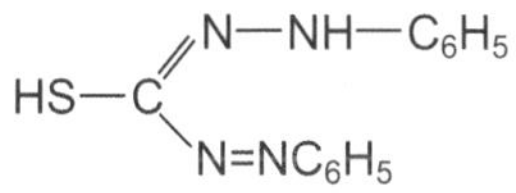

Dithizone

(*iii*) Chromium

Chromium salts are used in industrial processes. These enter the water bodies through discharge of waters. Chromium compounds are often added to cooling water for controlling corrosion. The level of Cr (VI) in drinking water is 3 to 40 ppb, the permissible level being 50 ppb.

Chromium is usually tested as chromate. If it is present as chromic ion (reduced form) it is first oxidised by placing a drop of the test solution on a filter paper and then a drop of 20 per cent NaOH solution and holding in bromine vapour. Then spot the paper with a drop of 5 per cent $AgNO_3$ solution strongly acidified with acetic and a red colour indicates chromium.

Analytical Procedure for Estimation of Chromium

Chromium is estimated by treatment with diphenylcarbazide in acidic medium at give red-violet coloured complex, which is measured spectroscopically. The present Mo, V, Fe and Cu interfere in the estimation. However, these can be removed by extraction with cupferron.

(*iv*) Copper

Copper is one of essential elements for humans. Salts of copper control biological growths in drinking water. The corrosion of pipe fittings, which contain copper alloys are responsible for the contamination of copper into drinking water. Excessive amount of copper are harmful.

Analytical Procedure for Estimation of Copper

Copper reacts with 2, 9-dimethyl-1, 10-phenanthroline (neocuproine) in neutral or weekly acidic solution (pH 3-9) to form a complex which is extractable into $CHCl_3$–CH_3OH mixture to produce a yellow solution. The latter can be measured spectrophotometrically at 457nm. This method is specific for copper. Beers law is valid up to 0.2mg Cu/25ml.

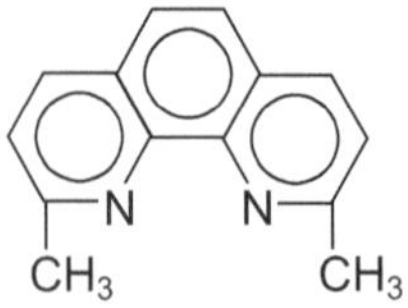

(Neocuproine)

(*v*) Fluoride

A fluoride level of 1ppm prevents dental decay. At higher concentration (5ppm) it causes mottled teeth and bone damage. Due to increased tendency of fluoride in public water supplies, its accurate determination has gained importance.

Analytical Procedure for Estimation of Fluoride

Fluoride reacts with Alizarin S lake to form colourless ZrF_6^{2-} and the dye. The colour of the dye lake become progressively weak with increase in the amount of F^-.

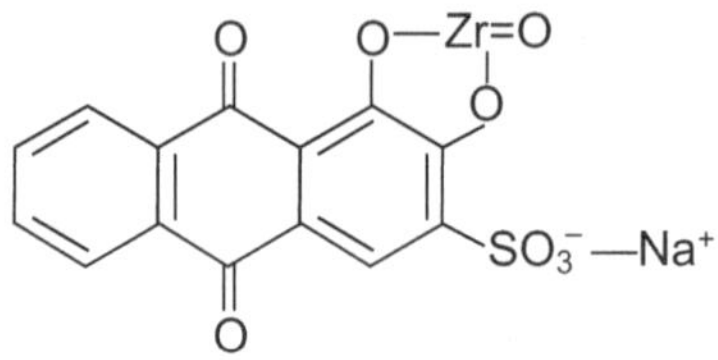

Zr-Alizarin Red S lake

(*vi*) Lead

Lead occurs abundantly in nature in the form of sulphide ore. During metallurgical operations, it finds its way into water bodies. Earlier lead was added to gasoline in the form of tetraethyl lead $[Pb(C_2H_5)_4]$ along with scavengers like 1, 2-dichloro- and 1, 2-dibromoethane. However, the use of leaded petrol has been banned in most of the countries and so is no longer considered to be the major source of lead in the atmosphere or water bodies. Lead pipes, which are used for transporting water is a minor source of lead in water bodies, since the lead pipes are coated with a layer of carbonate. However, even the small amount of lead in water bodies enters the food chain. The biochemical effect of lead is its interference with heme synthesis leading to haematological damage. Lead inhibits the aminolevulnic acid dehydrase enzyme. Due to this, the overall effect is the disruption of the synthesis of not only of haemoglobin but also of other respiratory pigments like cytochromes. Also, lead does not permit utilization of oxygen and glucose for life sustaining energy production.

Lead, being similar to Ca^{2+}, gets accumulated in the bones. Subsequently, it is remobilized along with phosphates from the bones which exert a toxic effect.

Lead poisoning can be cured by administration of chelating agents which strongly bind Pb^{2+}. One such example is administration of calcium chelate of EDTA. Pb^{2+} displaces Ca^{2+} from the chelate and the resulting Pb^{2+} chelate is excreted in the urine.

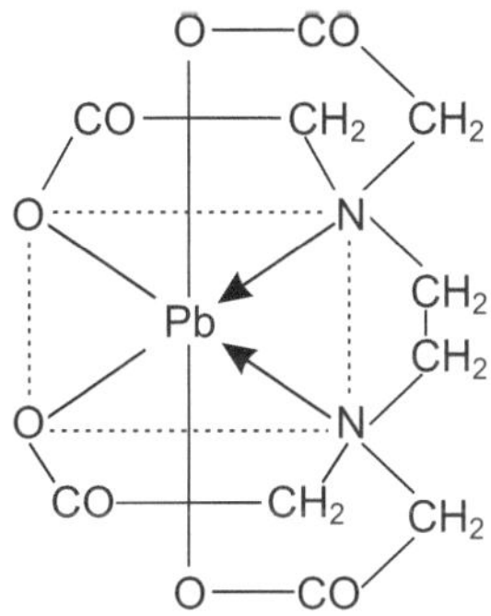

Lead-EDTA complex

Analytical Procedure for Estimation of Lead

The spectroscopic method is the most popular for analyzing environmental lead. In this procedure lead is extracted from the environment sample as a coloured lead-dithizonate complex. The colour intensity of the complex gives the concentration of lead.

The sample is which lead is to be determined (*e.g.* blood, tissue, soil etc.), is digested in a mixture of 80% HNO_3 and 20% $HClO_4$. By this process the organic compounds of the sample are oxidized to CO_2 and H_2O and Pb is converted into $PbNO_3$. The pH of the digested solution is adjusted to 8.5 – 9.0 by adding a buffer. Subsequently a mixture of ammonium citrate, potassium cyanide and hydroxylamine is added. Ammonium citrate prevents the precipitation of alkaline earth and transition metal hydroxides and phosphates. Potassium cyanide is a masking agent to prevent the extraction of metals, such as Cu and Zn. Hydroxylamine prevents the oxidation of dithizone.

A solution of dithizone in CCl_4 is then added. The lead of the sample is converted into a red lead dithizonate complex. This complex, being soluble in CCl_4 (the aqueous solution is discarded) is used to measure the absorbance at 510nm. Following the Lambert-Beer's Law, the intensity of absorbance is proportional to the concentration of lead-dithizonate, which, in turn, is proportional to the concentration of lead in the sample.

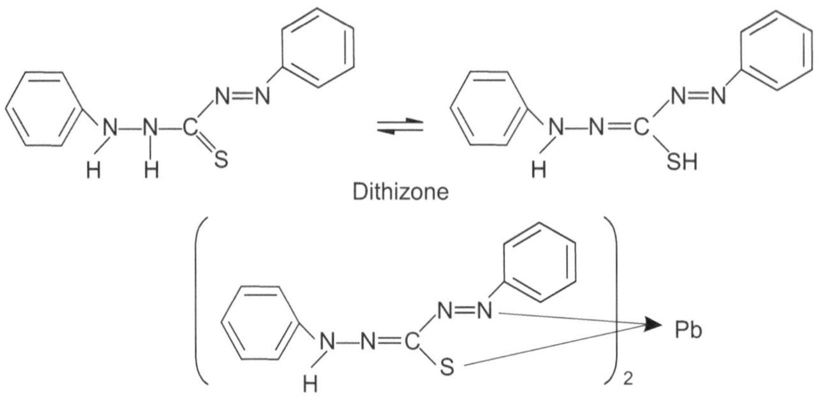

If lead is to be determined in the atmosphere, and then air is passed through a filter material, such as fiber glass or asbestos. The residue left on the filter is digested in HNO_3–$HClO_4$ mixture. In case lead is to be determined in water, a measured quantity of the sample is evaporated to dryness and the residue left is digested in HNO_3–$HClO_4$ mixture.

(*vii*) Mercury

Mercury is one of the most well known toxic metal. The toxicity or the toxic effects of mercury came to lime light in 1953–1960 in Japan by the incidence of Minamata Disease. At Minamata Bay in Japan, more than 100 people died and thousands were permanently paralysed by consuming fish contaminated with mercury. Even genetic defects were observed in some new born babies whose mothers had consumed contaminated fish from the Bay. The cause of the disease was contamination of water with mercury rich effluents discharged into the Bay by Chisso Chemical Company. It was discovered (1963) that the illness was due to methyl mercury poisoning, caused by eating contaminated fish from the bay. The Chisso plant used mercuric chloride catalyst, a non-toxic inorganic mercury compound in the production of acetaldehyde which was released in the industrial effluents. Subsequently, the sediments form the Minamata Bay was found to be rich in methyl mercury chloride. It was, therefore, clear that inorganic mercury underwent bio-methylation in aquatic system. It was also evident from the data obtained which showed the presence of methyl mercury chloride in concentrated form in the body of the fish.

It has been found that methylation of mercury in water bodies is brought about by a number of microorganisms called methanogenic bacteria. This process is called bio-methylation. The bio-methylation takes place under anaerobic conditions. The process is facilitated by the enrichment of water with organic impurities which permit growth of methanogenic bacteria; the bio-methylation proceeds smoothly in the pH range 5.5–6.5.

The Minamata incident was followed by a more tragic incident of mercury poisoning form Iraq (1972) where about 500 people died after consuming wheat which had been sprayed with mercury containing pesticide.

Sources of Mercury in Aquatic Systems

Mercury is discharged into the aquatic system by both natural and man-made sources. Amongst the natural sources is the ore of mercury, cinnabar (HgS) in igneous rocks. On an average, about 800 tonnes of mercury is released into the environment due to weathering of rocks. The lava, produced by volcanic eruptions is also a natural source of mercury contamination of environment. Due to this activity, the mercury content of lake Ontake in Japan has shown a steady rise due to repeated volcanic eruptions of the nearby Mount Kisoontake.

The mercury content in the aquatic systems has increased considerably due to human activities resulting from rapid industrialisation and urbanisation. Some of the human activities include the following:

- Widespread use of mercurial as insecticides, fungicides, bactericides and pharmaceuticals.
- Use of mercury amalgamated with silver, tin, cadmium and copper to make dental fillings.
- The chlor-alkali process using mercury cathodes is the single largest source of mercury in natural waters.
- Even after closing down various industries using or producing mercurial, the amount of mercury already accumulated in the beds of water bodies continues to be cause of major concern.
- Mercury vapour lamps, electrical switches, mercury batteries are the second largest source of mercury discharge in the environment.

  The impact of seed dressing is enormous. It causes widespread dispersal of mercury compounds. Also, mercury finds its way into the human food chain.

Toxic effects of mercury: The toxicity of mercury depends on its chemical species. Thus, elemental mercury (Hg) is non-toxic and quite inert. If swallowed, it is excreted without any serious harm. However, the vapours of mercury (it inhaled) are quite toxic. The vapours on inhalation enter the brain through blood stream causing damage to the central nervous system.

Mercurous ion (Hg_2^{2+}) forms insoluble chloride with chloride ions. Since the stomach in our body contains hydrochloric acid (chloride ions), it is not toxic. On the other hand, mercuric ion (Hg^{2+}) is quite toxic; due to its affinity for sulphur. It can attach itself to the sulphur containing amino acids of proteins. It also forms bonds with haemoglobin and serum albumin, both of which contain sulphydryl groups. Since this ion (Hg^{2+}) does not travel across biological membranes, it does not get access into biological cells.

The organomercurials particularly methyl mercury (CH_3Hg^+), which is fat soluble, are the most toxic species. The most harmful effect of RHg^+ is due to its ability to move through the placental barrier and enter foetal tissues. This is why the children born to mothers, who have mercury in their system are under developed; their central nervous system is damaged beyond repair. This is attributed to the deficiency of energy in the brain cells due to the presence of mercury. Mercury is also responsible for inhibiting the transport of sugars across the membranes.

Bioamplification of mercury: The organomercurial (e.g. methylmercury chloride) is incorporated in the food chains. Being soluble in the tissue of simple organisms, these penetrate inside the bodies of simple organisms. The complex species which feed on simple organisms accumulate the organo-mercurials in their system. This process is called bioamplication. Due to the process, organo-mercurials concentrate in the tissues of aquatic organisms, particularly fish. Bioamplication depends on the following parameters of water, viz. pH, temperature, hardness, salinity and contaminants. At lower pH (< 5), the uptake of methylmercury chloride is favoured. However, at pH more than 5, methylmercury chloride is converted into methyl mercury hydroxide (CH_3HgOH), which is insoluble in the lipid constituent of fatty tissues and so the bioamplication is reduced. The bio-amplification of methyl mercury chloride in fish increases with temperature and salinity. Bioamplication is reduced in hard water or in the presence of chelating agents like EDTA.

Effects of Mercury Poisoning

- Mercury inhibits enzyme action.
- Mercury reacts with sulphydryl groups in proteins and thus, disturbs the distribution of ions and interferes with the movement of fluids across the membrane.
- It is responsible for neurological disorders including dysphagia, ataxia, deafness and death.

Antidotes: The permissible limit of mercury in food is 0.5ppm. The level of mercury in blood should not exceed 1.0 ppm and in drinking water, it should not exceed 0.005 ppm. Some common chemotherapeutic agents for the treatment of mercury poisoning are given below:

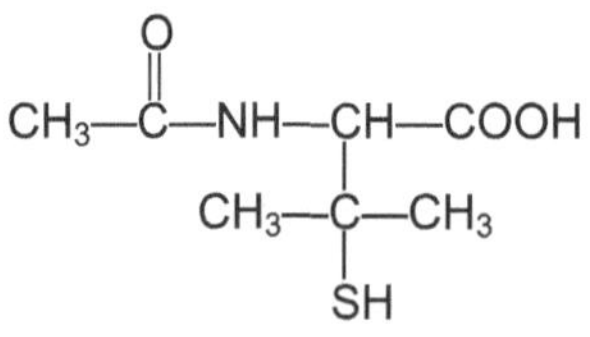

N-Acetyl-DL-penicillamine

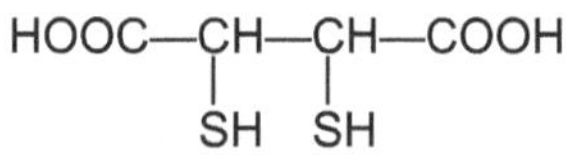

Dimercaptosuccinic acid

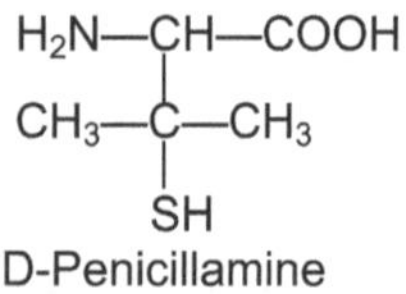

D-Penicillamine

These chemotherapeutic agents convert the intracellular methylmercury salts into a mobile complex that passes out of the body as urinary or faccal excretions.

Checking Environmental Pollution of Mercury: It has already been stated that mercury released into the environment by various agencies causes haveoc. In view of this, it is essential that the environment pollution due to mercury be checked. This can be achieved by the following procedures.

- All chlor-alkali plants must be stopped using mercury electrodes and switch to new technology. In fact, diaphragm cells are now being used.
- All alkyl mercury pesticides must be banned and replaced by less toxic substituents.
- Mercury-amalgam dental fillings must be banned.
- Release of mercury and its compounds as waste into water bodies should be curtailed by instituting recovery processes. Advance waste treatment can also contribute to reduction of pollution caused by mercury.

Using all possible precautions and new technologies, it has been possible to reduce the mercury pollution. The main problem the world is facing with it is that the water bodies whose underlying sediments are already overburdened with mercury will continue to release the metal for a considerable length of time. Attempts to dredge up the mercury containing sediments makes the problem worse since dredging stirs up the bottom level and mercury will spread in whole of the water bodies. It has been suggested to prevent the release of mercury from the sediments by covering it with a layer of ferrous sulphide and sand. However, this method is too cumbersome considering the vast oceans.

Analytical Procedure for estimation of mercury: Mercury is estimated by the reduction-aeration-flameless atomic absorption method. For determining mercury in water, rock or soil, the sample (if mercury is to be determined in water, the water sample is evaporated to dryness and the residue is used) is digested with H_2SO_4 or HNO_3. Any organic matter if present is oxidized with a mixture of H_2O_2 and $KMnO_4$. A mixture of NaCl-NH_2OH is added to destroy excess of oxidizing agents, white retaining mercury in the mercuric (Hg^{2+}) form. Subsequently a solution of stannous chloride is added to reduce mercuric ions to elemental mercury.

$$Sn^{2+} + Hg^{2+} \longrightarrow Sn^{4+} + Hg$$

The elemental mercury is volatilized by passing a stream of air at a rate of about 2 liters min^{-1}. The air stream is directed to an atomic absorption spectrophotometer. Intensity of the absorbance at 273 nm is directly proportional to the concentration of mercury in the air stream, which in turn is proportional to the total mercury concentration in the sample.

In case the sample contains mercury as methyl mercury (as is generally the case in sea waters), the procedure has to be modified. Methylmercury is not reduced by stannous chloride. However, a mixture of stannous chloride and cadmium chloride is effective in converting methylated mercury to elemental form. The elemental mercury is carried to the atomic absorption spectrophotometer by an air stream and determined as already described.

(*viii*) Selenium

Selenium is essential at low level but toxic at higher levels. It is a suspected carcinogenic agent. It finds its way into water bodies from natural geological sources, sulphur and coal. The selenium content of drinking water is less than 10 ppb, which is the permissible limit.

Analytical Procedure for Estimation of Selenium

Selenium is liberated from its solution as selenium bromide by distillation with HBr and Br_2. The selenium bromide in aqueous solution becomes selenous acid. The excess bromide is reduced by treating the solution with SO_2. Selenous acid is reduced by NH_2OH. HCl to selenium, which is precipitated as stable red colloidal suspension and can be determind spectrophotometrically.

Alternatively, selenium is first oxidized with acid $KMnO_4$ to SeO_4^{2-} and then reduced to SeO_3^{2-} in warm 4N HCl. Then it is reacted with diaminobenzidine at pH 1.5 to form a chelate, which is extracted into toluene and the yellow extract is measured spectrophotometrically at 420 nm.

2.9 WATER QUALITY STANDARDS

Depending on the use for which water is intended to be used it should have the right parameter. For detail see Tables 2.14 and 2.15

2.10 SOME OTHER INGREDIENTS IN DRINKING WATER

Besides the dissolved inorganic impurities (section 2.7) and toxic metals (section 2.8), the water may contain impurities of chloride, fluoride, sulphate, sulphide, nitrate, nitrite, cyanide, etc. (see Table 2.15). Some of these are described below:

(*i*) Chloride

In drinking water, chlorine is relatively harmless if present in amount below 250 ppm. In waste water the chloride content is higher than in raw water. However, a high chloride content in water bodies is harmful to metallic pipes and also to agricultural crops.

Chlorine is simply determined by titration with $AgNO_3$ solution using $K_2Cr_2O_4$ as an indicator. The end point is indicated by the appearance of permanent reddish tinge (mohr's method). The method is valid for 0.15 to 10mg. chloride. (1 mL of 0.0282N $AgNO_3 \equiv 1$ mg Cl^-). For smaller quantities, potentionetric titration with $AgNO_3$ solution using glass and Ag-AgCl electrode system is used.

(*ii*) Residual Chlorine

The chlorine left unreacted in water after chlorination is referred to as residual chlorine. The effectiveness of chlorination of raw water for public water supply can be checked by estimation for free or residual chlorine in water samples.

In fact free available or residual chlorine consists of Cl_2, HOCl and $HOCl_2$. Free Cl_2 reacts instantly with N, N-diethyl phenylenediamine (DPD) indicator to form a red colour. This colour is discharged on addition of $(NH_4)_2$ $Fe(SO_4)_2$, ammonium iron (II) sulphate solution.

The procedure consist in taking a known volume (100mL) in a conical flask, adding DPD reagent (5mL, 0.1 g.N, N-diethyl-1, 4-phenylenedia-mmonium sulphate and 2mL of 2N H_2SO_4 and 2.5 mL of 0.02 MEDTA made to 100 mL) and phosphate butter (5mL obtained from 46 g. KH_2PO_4, 24g.Na_3PO_4. $12H_2O$/L). The resulting solution is titrated with ammonium iron (II) sulphate solution (1.1 g in 1 litre containing 2 mL $2NH_2SO_4$) till the solution become colourless or a permanent faint pink.

$$1 \text{ mL of } (NH_4)_2 \text{ Fe (II)}(SO_4)_2 \equiv 0.1 \text{ mg free chorine.}$$

Free chlorine above 40pbb can be conveniently estimated iodometrically. At pH 8 or less free Cl liberates I_2 from KI, which can be titrated with standard 0.02N $Na_2S_2O_3$ solution.

(*iii*) Chlorine Demand

Chlorination is necessary for polluted waters such as sewage, trade wastes, industrial wastes and water highly contaminated from organic sources. This is helpful in controlling the odour of water and bacterial growth. A measure of the chlorine demand is required to estimate the quantity of chlorine necessary to add to a given volume of water.

Chlorine demand is defined as the quantity of chlorine required to produce residual chlorine content between a trace and 0.1 ppm after a contact of 10 minutes. The measurement of chlorine demand helps in estimating the quantity of chlorine necessary to add to a given volume of water.

The procedure for determining the chlorine demand is given below:

Take 250mL of the water sample in a reagent bottle. Add to it standard chlorine water (containing 0.5 mg of chlorine) from a burette. Stir the mixture, withdraw 2.5 mL of the solution with a pipette on a spot plate and test immediately for free chlorine with a drop of *o*-toluidine reagent (if free chlorine is present it gives a light yellow colour). Continue adding 0.5 mL portions of standard chlorine water, mixing and testing until there is a slight excess of chlorine. The quantity of chlorine required is regarded as the immediate chlorine demand.

Now again take a similar sample, add to it standard chlorine water in an amount equivalent to the immediate chlorine demand plus 0.5mL more. Shake the mixture well for exactly 10min. After 10 min. immediately test a portion for residual chlorine as described above.

$$\text{The chlorine demand} = \begin{array}{c}\text{mL of standard}\\ \text{chlorine water}\end{array} . \begin{array}{c}\text{grams chlorine per}\\ \text{litre in standard}\\ \text{chlorine water}\end{array} . \frac{1000}{25} \text{ ppm.}$$

(*iv*) Fluoride

The permissiable limit of fluoride in drinking water 0.5 mg/L. Excess of fluoride is harmful. For details see page 46 and 62.

(*v*) Sulphate

The main source of sulphate in natural water is due to drainage of wastes from mines. Mine drainage wastes may contribute high sulphate due to pyrite oxidation. Excess Na_2SO_4 should not be present in drinking water as they cause cathartic action.

Sulphates is precipitated as $BaSO_4$ by the addition is $BaCl_2$. The precipitation is carried out near the boiling temperature. The resultant matter is digested, the precipitate filtered, washed with water till free of Cl, ignited and weighed.

(*vi*) Sulphide

Ground water generally contains sulphide. Hot springs and common wastes waters coming from the decomposition of organic matter, from industrial wastes or from reduction of sulphides are common sources of sulphide. H_2S gas gets oxidized biologically to H_2SO_4 in the pipe wall.

Sulphide is estimated by taking 500 mL of the water sample in one litre flask and adding 1 mL of 2 N zinc acetate (200 g/L) and 2 mL of 1 N NaOH. To the above solution is added 20 mL of 18 N H_2SO_4 and rapidly distilled the solution into 200 ml of 2.2% zinc acetate solution, when ZnS gets precipitated. To the receiving flask containing ZnS is added 10 mL of 0.05 N I_2 solution after acidifying the solution with H_3PO_4 (124 mL/500 mL distilled water). After 25–30 minutes, the excess of iodine is back titrated with 0.05 N sodium thiosulphate solution using starch as an indicator.

$$H_2S + I_2 \rightleftharpoons 2H^+ + 2I^- + S$$

$$1 \text{ mL of } 0.05 \text{ N} I_2 \equiv 0.85 \text{ mg } H_2S$$

(*v*) Nitrate and Nitrite

Both nitrate and nitrite are deemed to be potentially hazardous to health if their maximum administrable concentrations of 50mg L^{-1} and 0.1 mg L^{-1} resp. are exceeded. Nitrate is particularly dangerous to infants less than six months old, causing a disease called methemoglobinemia. A limit of 10mg L^{-1} for nitrate has been proposed for drinking water (see also page 51 and table 2.15).

Nitrates occur in traces in surface water, but more in ground water. In the oxidation of NH_3 to NO_2 and in the reduction of NO_3, which occurs in waste water treatment plants and natural waters, nitrites are an intermediate product.

Total nitrate and nitrite are estimated by reducing them to NH_3 by Devarda's alloy (Cu : Al : Zn 50 : 45 : 5) in strongly alkaline solution. The liberated NH_3 is distilled into excess of standard alkali and finally estimated spectrophotometrically or titrimetrically.

$$3NO^-_3 + 8Al + 50H^- + 2H_2O \longrightarrow 8AlO_2^- + 3NH_3$$

In the titrimetrical method, the excess of alkali is back titrated with standard acid using methyl red as indicator.

$$1 \text{ mL } H_2SO_4 = 0.06201 \text{ g. } NO_3^-$$

In the spectrometric method, a known volume of the solution of NH_3 in H_2SO_4 is neutralised to pH 4.5, Neslers reagent (2 mL) added and the absorbance is estimated at 424 nm. The spectrophotometric method is valid for NO_3 concentrations greater than 0.5 ppm, while for concentrations exceeding 5 ppm, titrimetric method is followed.

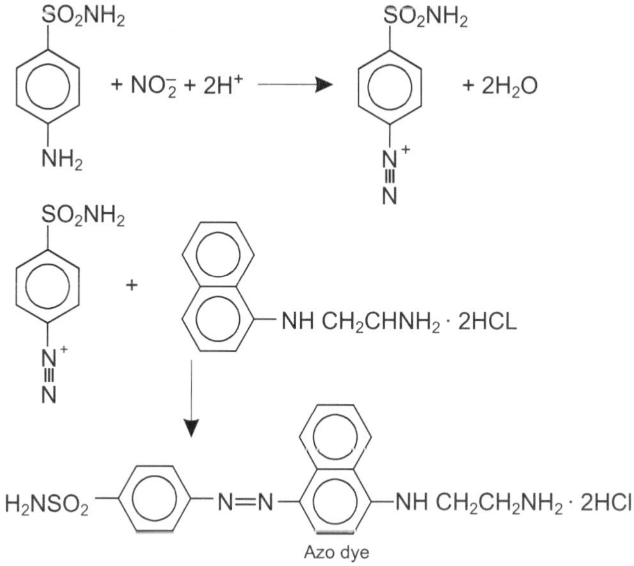

The nitrite is estimated by taking a sample of water (say 40 mL) in a volumetric flask. The pH is adjusted to 7.0 and sulfanilamide solution (2 mL 50 g. in 500 mL of 1.2 N HCl) added. The resulting solution is shaken and allowed to stand for 10 min. Subsequently a solution of N–(1 naphthylethylenediamine dihydrochloride (2 mL, 0.83 g in 200 mL warm water, solution cooled, filtered and diluted to 250 mL with glacial acetic acid) is added, the solution diluted to 50mL, mixed throughly. The formed purple azo dye is measured within 2 hrs at 543nm. Measure against standards covering the range of NO^-_2 1–25 ug l^{-1} (at the final dilution).

(*vi*) Cyanide

Free cyanide (CN^- or HCN) is a toxic and hazardous water pollutant. It usually comes in water bodies from metal refining and cleaning, electroplating and other industrial process. The CN^- is destroyed by oxidising agents like Cl_2 during storage. For estimation of CN^-, sulphides if present are removed by precipitation with 10 per cent Cd $(NO_3)_2$ solution and rejecting the precipitate and filering of. The filtrate is used for estimating CN^-. For this 250 – 500 mL sample is taken in a distillation flask in a fume chamber and refluxed for one hr. with H_2SO_4 (50mL, 18N) and $MgCl_2 6H_2O$ solution (20mL, 50%). The evolved HCN gas is collected in 500mL of N. NaOH. Finally the cyanide content is measured titrimetrically or spectrophotometrically.

In the titration method, 0.5mL of 0.02 per cent *p*-dimethylamino-benzal rhodanine solution in acetone is added to an aliquot of the distillate and then titrated against 0.02 N $AgNO_3$ solution till the colour changes from yellow to salmon blue. It is compared against a blank under as nearly as possible identical experimental conditions.

In the spectrophotometric method, the CN^- in the alkaline distillate is converted into CNCl by addition of chloramines T at pH 8.0. To the solution is added pyridine-barbituric acid and the blue dye thus formed is measured at an absorbance of 578 per cent nm. The pyridine-barbituric reagent is prepared by mixing 75mL pyridine, 15g barbituric acid and 15mL conc.HCl, diluted to 250mL.

(*vii*) Oils

The oils in a water sample is estimated by taking 2 liters of thoroughly mixed water sample in a beaker and evaporating the solution to dryness on a steam bath. The residual oil is extracted with petroleum utter, and the clear solvent evaporated on a steam bath. The residual oil is weighed. The amount of oil in ppm = weight of oil × $10^6/2000$.

2.11 PURIFICATION OF WATER

2.11.1 Introduction

During the past few decades due to rapidly developing technology, increase in population and urbanisation, there has been alarming rise in the pollution of water all over the world. The pollution of water bodies has become a universal problem. It is the matter of ever increasing concern to mankind. All inhabitants must have a continuous and plentiful supply of water for drinking, cooking, washing and sanitary purposes i.e. domestic and public uses. Water should be safe and palatable.

Water for drinking is mostly obtained from water bodies like rivers, lakes etc. These water bodies are polluted by the discharge of domestic wastes, untreated sewage and industrial effluents from various industries like textile industry, paper and pulp industry, leather tanning industry, agrochemical industry, petrochemical industry, food processing industries, diary industry, soaps and detergents and industries dealing with chemicals. At times water bodies are also polluted by hazardous wastes including cyanide wastes, metal finishing wastes, heavy metals including mercury, arsenic, thallium and cadmium, wastes from paints, pigments, printing inks, dyes and asbestos. Biological agents like bacteria, parasitic fungi and protozoa arising from human sewage, animal and plant wastes and decaying organic matter also pollutes water bodies. In case the wastes are discharged on the land, these pollute the underground water by seepage.

Several incidents of improper disposal of wastes have been reported. A typical episode occurred in Japan (Minamata) due to discharge of industrial wastes containing mercury.

Wastes which are non-biodegradable have the worst effect on human health. Such wastes include DDT, plastics, polymers and fluorinated compounds. These wastes persist for longer times.

The word sewage is used for the combined liquid wastes discharged from all domestic (toilets etc.) and also industrial sources. It contains faecal matter, urine soap and detergents. This waste is referred to as municipal waste. On an average about 75 per cent of water pollution is caused by sewage wastes. These wastes have normally high turbidity appearing greyish (some become black due to partial decomposition). These wastes develop a characteristic odour of rotten eggs due to hydrogen sulphide. The sewage contains inorganic matter like nitrates, phosphates, sulphates, chlorides of calcium and magnesium. Besides, these also contain some trace quantities of toxic metals like copper, chromium, zinc, manganese, lead and nickel. The common organic matter in sewage include fatty acids esters, amino acids, amides, amino sugars, proteins and their decomposition products.

There has been during the past years, uncontrolled dumping of wastes from rural areas, towns and cities into the water bodies. Tremendous quantities of sewage are produced due to extensive population increase. Much of the sewage does not receive any treatment before it is discharged into the water bodies. So, such water cannot be directly used for any specific purpose and is highly harmful and toxic. A number of hazardous and toxic chemicals and pathogens enter the sewage system. Most industrial effluents are not easily degradable. We have discussed earlier (section 2.5.4) the impact of water pollution on human health.

2.11.2 Treatment of Water

The procedure used for treatment of wastewater depends on the source from which it has originated. The only way to check the pollution caused by various industries is that each industrial unit must process its own wastewater. The processed water can be reused in the same industry.

Waste water, particular municipal wastes (including domestic and household wastes) is treated in the following three stages

- Primary or preliminary treatment
- Secondary treatment or biological oxidation process
- Tertiary treatment or complete chemical treatment

In the Primary treatment, the raw sewage water is first passed through a series of screens to remove large floating material. All non-biodegradable and chemical non-degradable solids like rocks, sand, grit, plastic and metal parts are separated by this method. The removal of these solids allows the polluted water flow freely. The water is then passed through a chamber (called the grit chamber), which is packed with sand and small stones. This is helpful in removing the dirt, which is disposed off. The resulting filtered water then enters a sedimentation tank, in which particulate matter settles down to form the sludge. Sometimes, it is necessary to add certain chemicals like alum, which help to hasten the settling process. The sludge is removed and used after processing as manure. The above primary treatment removes about 30-40 per cent of the pollutants in the form of suspended matter.

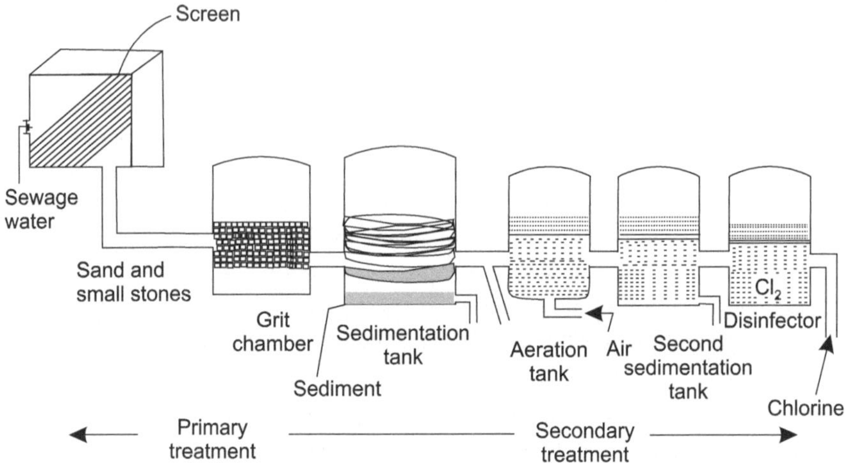

Fig. 2.3 Sewage water treatment

In the secondary treatment, the oxidizable biodegradable solids are removed with the aid of living organisms. In this procedure, the water from the primary treatment, viz. from the sedimentation tank is taken to an aeration tank in which air is pumped along with some sludge from the final (second) sedimentation tank. The sludge contains aerobic (oxygen rich environment) bacteria that break down organic pollutants in the water. After several hours in the aeration tank, the water is taken to the second or final sedimentation tank, in which most of the sludge settles dam. The total sludge is subjected to microbial digestion, in which process methene gas is evolved, which is used as a fuel. The water from the final sedimentation tank is then disinfected in disinfection tanks usually by chlorine; this removes the disease-causing organisms.

Normally, primary (or preliminary treatment) and secondary treatment (or biological oxidation process) is the complete treatment of water (see Fig. 2.3)

Tertiary treatment is the final step in the purification of water. It includes purification of water by treatment with chemicals. This is helpful to remove highly soluble salts as precipitates by the addition of alum or other flocculating agents. Such treated water still contains nitrates, nitrites, phosphates and other non-biodegradable chemicals (like herbicides and insecticides) and biological agents (like bacteria, virus etc). Such water is rendered alkaline by the addition of time, which precipitated the phosphates. Subsequently some metallic hydroxides are added. By this treatment phosphates are precipitated as insoluble phosphate salts.

Nitrates and nitrites, which are very soluble in water, are removed by electrodialysis. Chemical coagulants like alum, ferrous sulphate, ferric sulphate etc. also help to remove nitrates and nitrites. Aeration of resulting water helps to remove foul odour and tastes. Finally, water is treated with chlorine or ozone, both chlorine and ozone are toxic to bacteria and virus.

Complete removal of pollutants can be affected by advanced procedures like reverse osmosis, electrodialysis and vigorous agitation with air. This treatment does not require any chemical. UV rays are helpful to kill most of the bacteria in water.

2.12 CONSERVATION OF WATER

There has been a considerable increase in the consumption a water over the years. It is therefore in the interest of mankind to conserve water. Water conservation is the careful use and protection of water resources. Maximum amount of water is used in agriculture, for domestic needs and industries.

(*i*) Agriculture Use

Agriculture in the largest user of water. About 80% of the total fresh water is used for this purpose. There could be tremendous saving by using the following procedures.

- Reduce seepage and evaporation of the water by using lined or covered canals.
- Use computer monitoring and scheduled release for maximum efficiency.
- Irrigate when evaporation loss in minimum, for example at night or early morning.
- If it is possible, use much plant around to retain water.
- Grow crops that need less water.

(*ii*) Domestic use

Nearly 10 per cent of the total fresh water is used for domestic purposes. This can be considerably reduced by the following practice.

- Use more efficient bathroom fixtures, such as low flush toilets that required 1.5 gal per flush rather than the standard 5 gal per flush toilet.
- Use showerheads that deliver less but sufficient water.
- Turn off water when not needed for washing, brushing teeth, shaving etc.
- Flush the toilet only when necessary
- All leaks should be fixed at the earliest
- Using dish washers or washing machines can reduce water consumption
- Water lawns and plants in early morning, or at night to reduce evaporation loss.
- Use drip irrigation and place water holding mulch around garden plants.

(*iii*) Industrial use

The use of water can be decreased by restoring to in-plant treatment and recycling of water.

EXERCISE

1. Explain why water in called universal solvent.
2. What is the main difference between the chemical composition of ground water, river water and sea water?
3. What are wetlands?
4. Explain and illustrate hydrological cycle.
5. Explain what you understand by aquatic pollution.
6. Discuss some important sources of water pollution.
7. How will you deal with marine pollution?
8. What is the effect of aquatic pollution on human health?
9. What are the dangers if drinking water is contaminated with nitrates.
10. Write a note on Eutrophication.
11. Explain the terms DO, BOD and COD. How are these determined?
12. What is the effect of contamination of sea water with cadmium and mercury?
13. Explain Bioamplication of mercury.
14. Write a note on chlorine demand.
15. How is water purified for drinking purposes?
16. Write a note on conservation of water.

Chapter 3

Soils

3.1 INTRODUCTION

Soil is the top layer of the earth on which plants and animals grow. It is defined as the layer of reconsolidated particles derived from weathered rock, organic matter (humus), water and air and form the upper surface of the earth and supports plant growth. Soil plays an important role since it produces food materials on which the survival of mankind depends. It is also a vital component of the environmental Biogeochemical cycles. The formation of soil depends on the parent material (the original material from which soil is derived), the climate and the topography of the area, the organisms present in the soil, and the time in which the soil has been developing

The process of breakdown and alteration of rocks on the earth's surface by mechanical or chemical processes is called weathering. Mechanical (physical) weathering is the splitting of rocks through the action of heat and extreme temperature changes. Physical weathering elements originate from climate. e.g. temperature, ice, gravity and wind. Due to sun's heat, a rock becomes warm, which makes it to expand. During night the rock cools more quickly from outside than form inside. This results in the development of cracks in the rock. The cracked layer of rock breaks into small fragments as pieces. Water may run into cracks, which freeze with the change in temperature. The freezing of water causes the cracks to become deeper. Chemical weathering includes solution (the dissolving of solid materials by water); Carl ovation (the dissolving of soluble rocks and minerals by a weak carbonic and formed by the combination of water with atmospheric carbon dioxide); oxidation (the combination of atmospheric oxygen with rock materials); and hydration (the chemical combination of rock material). Organic or biological weathering may involve both chemical mechanical processes caused by both plants and animals e.g. burrowing animals and plant roots may physically break up rocks, lichens, which can grow on bare rock surfaces, cause decomposition. Also lichens, have acid present in them which cause chemical changes and make the rocks crumble. When glaciars move, the crumbled rocks are pushed ahead of them. This pushing help the rock to crumble. Water and chemicals present in the water, dissolve and wear away some kind of rocks. Once this process starts the remaining rock crumbles. The land slides due to earthquakes are also responsible for fragmentation of rocks by forces of impact and aversion. The wind help in carrying small fragmented particles rocks to the surface of the earth.

The rocks are subjected to a continuous process of physical and biological weathering. Along with physical weathering, chemical weathering also occurs by water, oxygen carbon dioxide. Water soluble minerals like lime-stone, gypsum etc. from weathered rocks increase the presence of CO_2 and organic acids formed as a result of decay of remains of animals and plants. Water is also responsible for hydrolysis along with CO_2. Thus the silicates weathered rocks are hydrolysed.

$$K_2Al_2Si_6O_{16} + 2H_2O + CO_2 \longrightarrow Al_2O_3\,2SiO_2 \cdot 2H_2O + 4SiO_2 + K_2CO_3$$

This process releases Ca, Mg, K, Na and silicates in the soil. All these are responsible for increasing the fertility of soil. The CO_2 and H_2O combine to from carbonic acid (H_2CO_3) which is also responsible for the weathering of rocks.

3.2 SOIL PROFILE

Soils are classified in terms of their structure and texture. The structure of a soil is the way in which the individual soil particles are bound together to from aggregates or pads. The texture of a soil denotes the proportion of various particle sizes that it contains.

A number of distant horizontal layers often seen in a vertical section (profile) of soil are known as **soil horizons**.

These different soil horizons can be seen in Fig. 3.1. There are four basic horizonscare common to most soils.

The top layer which is the uppermost 'A horizon (or topsoil) contains organic matter and is very important for vegetation & agricultural crops. Top soils is lost due to deforestation over the years It takes decades to from even a minor thickness of top soil.

The next layer is the (subsoil) B horizon, which contains little organic matter.

The underlying is the C horizon, consisting of weathered rocks from which soil originates. The lowest layer of rocks is the hared zone called bed rocks.

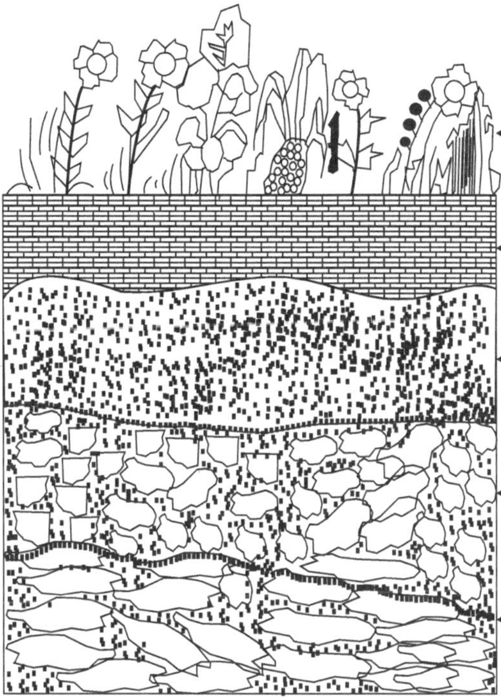

Fig. 3.1 Soil profile

3.2.1 Soil Composition

The soil is comprised of inorganic (90–95 per cent) and organic matter (5–10 per cent). Besides these soil contains water and air. The composition of the soil varies considerable depending on the type of soil.

Inorganic Components of Soil

Most of the inorganic components of soil exist as silicate minerals. The common elements in the soil are oxygen (46.6 per cent), silicon (27.7 per cent), aluminium (8.1 per cent), iron (5.0 per cent), calcium (3.6 per cent), sodium (2.8 per cent), potassium (2.6 per cent) and magnesium (2.1 per cent). Commonly found silicates in the soil are quartz (SiO_2), Orthoclase ($KAlSi_3O_8$), albite ($NaAlSi_3O_8$) and epidote [$4CaO.3(AlFe)_2\ O_3.6SiO_2.H_2O$]. Many soils contain iron oxides, FeO (OH), magnetite (Fe_3O_4). Certain soils contain titanium, and manganese oxides. Calcium carbonate is another common component of soil.

Soil also contains a mixture of minerals as carbonates, sulphates, chlorides and nitrates. Some soils also contain clay mineral, which are hydrated iron and alumium silicates. These clay minerals bind cations such as Ca^{2+}, Mg^{2+}, Na^+, K^+, NH_4^+, so that these cations are not leached by water and are available to plants as plant nutrients.

Organic Compounds in Soil

The fertility of the soil depends on the organic matter. The organic matter constitutes 5–10 per cent of the soil. The major organic constituents in the soil are humus, lipids, saccharides and nitrogen and sulphur containing compounds.

1. **Humus.** The humic materials constitute the most important component. These form complexes will a number of cations, so that these are retained in the soil. In general, the humic substances are high molecular weight macromolecules. These are obtained in soil as degradation resistant residues from the decay of plant and dead animals.
2. **Lipids.** These constitute fats, resins and waxes and are extractable by organic compounds. The physical properties of soil depends on these compounds.
3. **Saccharides.** Examples include cellulose, starches, hemicelluloses, etc. These are the major food materials for soil microorganisms and help to stabilise the soil aggregates.
4. **Nitrogen containing organics.** Examples include amino acids, amino sugar etc. These provide nitrogen to the soil for fertility.
5. **Phosphorus containing compounds.** Examples include phosphate esters, phospholipids, etc. These are the source of phosphorus in the soil.

Water and Air in Soil

The main source of water in the soil is through dew, rain, snow, hail, irrigation etc. Water is essential for the growth of plants and is an important medium for transporting plant nutrients from soil into the roots and other parts of the plants. Any excess of water from the plants finds its way into the atmosphere by the process of transpiration. In the soil, water stands at a certain level known as **water table**. In case the water table in too low, the plants are unable to get the required amount of water. On the other hand, if the water table is too high, the plants drown. Rice is the only crop which grows on water logged soils.

The air in the soil contains oxygen, nitrogen and carbon dioxide. The decay of organic matter in the soil consumes O_2 and produces CO_2. The result is that soil contains less O_2 (about 15 per cent compared to 21 per cent in the atmosphere) and more CO_2.

Macro and Micro Nutrients in Soil

Plant require various nutrients for their growth. These nutrients are derived by plants from the soil. The nutrients that are needed in larger amounts are called **macronutrients**, and those that are needed in trace amounts are called **micronutrients**.

The macronutrients that are present in the soil are carbon, hydrogen, oxygen, nitrogen, phosphorus, calcium, copper, iron, magnesium, sulphur, sodium and potassium. The first three nutrients, i.e. carbon, hydrogen and oxygen are obtained from the atmosphere and water. Nitrogen may be obtained by some plants directly from the air with the help of nitrogen fixing bacteria. Nitrogen, phosphorus and potassium are generally added to the soil as fertilizers (like ammonium sulphate, urea, potassium nitrate, super phosphate and NPK). The other macronutrients are present in the soil.

Some soil may become deficient in calcium due to up take by plants, leaching by carbonic acid, etc. Calcium deficient soils are treated with lime to provide calcium to the soil/plants. The latter process is called **liming**.

The essential micronutrients required by the plants are aluminium, boron, bromine, zinc, cobalt, manganese, zinc, vanadium and molybdenum. These elements are required in trace amount. Excess of these elements has a toxic effect.

From the soil, the nutrients find their way into plants. When the plants/crops are consumed by humans, these nutrients are used in various life processes. Most of the micronutrients serve as components of essential enzymes. When the plants, animal and human being die, the nutrients return to the soil and in this way most of this elements are recycled.

3.3 SOIL POLLUTION

Soil has been used as a dumping ground for most of the waste products–domestic, human, animal, industrial and agricultural. Every year the solid wastes dumped into the soils are increasing at an alarming rate all over the world. Huge amount of a number of hazardous chemicals and several thousand tons of wastes are ultimately dumped on the land. These are leached by municipal and industrial wastes and are responsible for pollution of ground water (also see Section 2.3.1). The problem of soil pollution is compounded by the use of agrochemicals (e.g. fertilizers, pesticides, fungicides, bactericides, insecticides, etc). Besides these, the soil is polluted by deadly pathogenic compounds. The soil pollution differ from air and water pollution in the sense that the pollutant in soil remain in direct contact with soil for relatively longer periods. In fact, the soil has been heavily polluted as a result of industrial revolution and green revolution. The humans are basically responsible for the pollution of land.

3.3.1 Sources of Soil Pollution

The main sources of soil pollution are

- Industrial Pollution
- Urban and Domestic Wastes
- Radioactive Pollutants
- Agrochemicals
- Plastic and Polymers
- Chemicals and Metallic Pollutants
- Biological Agents
- Acid Rain

Industrial Pollution

The major cause of soil or land pollution are the wastes, by-products generated from various industries. These wastes are discharged from various textile industries, paper and pulp industries, leather and tanning industry, petrochemical industry, food processing industry, soap and detergent industries dairy farms and metal processing industries etc.

It is believed that in these industries 50 per cent of the raw materials become a part of the waste products. These pollute not only the land or soil but they also pollute water bodies.

Urban and Domestic Wastes

The urban and domestic wastes together are referred to a municipal wastes. They constitute garbage, and waste materials like plastics, glass, metallic containers, fibers, papers and a number of discarded products. It is estimated that in UK and India about 4,00,000 tonnes of solid waste is produced daily. The situation is so alarming that the land fills sites used for more than 25 years cannot take any more refuse.

A sizeable portion of urban and domestic wastes are not biodegradable and so are very harmful. For example, certain items like paints, varnishes, oils etc. create the worst type of soil pollution problems. The leaches from the dumping sites and sewage along with industrial effluents are extremely harmful and dangerous and pollute even the ground water. These leaches contain poisonous gases along with partially decomposed materials from food remains, toxic chemicals and pathogenic bacteria which can be very harmful and may cause various diseases.

Radioactive Pollutants

Radioactive wastes (nuclear wastes) are any solid, liquid or gaseous waste materials that contains radio nuclides. These wastes are produced in the mining and processing of radioactive ores, the normal mining of nuclear power stations and other reactors; the manufacture of nuclear weapons; and in hospitels and research laboratories. Radioactive pollutants may be highly dangerous to all living matter because they may contain radionuclides having half-lifes of many thousands of years, their disposal has to be coutrolled with great stringency.

In soils, the radioactive pollutants come from the following sources.

(*i*) **Radioactive Fallouts.** These result from testing of nuclear weapons. Considerable amounts of fission products, particularly C-14, Sr-90, I-131, CS-137, etc. are dispersed in the atmosphere as a result of nuclear weapon testing. These nuclides spread all over the globe and ultimately settle on earth's surface as radioactive fallouts, from where these find their way into the food chain. Most of the nations of the world are in favour of complete Nuclear Test Ban Treaty. However, so far, only few countries have signed this treaty

(*ii*) **Radio-isotopes.** These are used in research, industry and medicine and are also generated during the operation of nuclear reactors for generation of power and the nuclear fuel processing plants (see Chapter 5).

(*iii*) **Natural Radioactive Nuclides.** The earth's crust contains some radioactive materials like U-238, Th-234, Rh-226, which continually emit radiations. The food crops grown in soil contain radioactive nuclides and the drinking water percolating through the soils also contain some radioactive nuclides like potassium -40, carbon-12 and Rn-222. These enter the body through the food chain.

(*iv*) **Accidents in Nuclear Reactors.** These are rare, but if, take place, they become the major cause of emission of radio isotopes or nucleotides into the air. Typical disaterous nuclear accidents are the Three Mile Island nuclear accidents (USA, 1979) and Chernobyl disaster (USSR, 1986).

Agrochemicals

Agrochemicals are the chemicals used in farming, especially for killing insects or for making plants grow better. They are also used in agriculture to increase the yields. They include fertilizers and pesticides.

Fertilizers

These are used to increase the fertility of soil, which is essential for better yields of the crops. Fertilizers are of two types, viz., natural and synthetic. A typical example of natural fertilizers is composts. Among the synthetic fertilizers are the nitrates, phosphates etc. of sodium and potassium use of excess of fertilizers result in washing down of these into lakes by rain; this in turn is responsible for eutrophication of the lakes.

Broadly speaking, fertilizers are of three types:

- Nitrogen based fertilizers like sodium nitrate ($NaNO_3$), ammonium sulphate, $(NH_4)_2$ SO_4 and urea, NH_2CONH_2.
- Phosphate based fertilizers like super phosphate [$Ca_2H_2(PO_4)_2$. $CaSO_4.2H_2O$], triple super phosphate [$Ca_2H_2(PO_4)_2$. $CaHP_4$].
- Potassium based fertilizers like potassium chloricle (KCl)and potassium sulphate (K_2SO_4)

Excessive use of fertilizers or repeated use of the same fertilizer pollutes the soil. For example, when ammonium sulphate is repeatedly used as a soil fertilizer, the NH_4^+ ions are used up by successive crops but SO_4^{2-} ions get accumulated in the soil. Being acidic, these SO_4^{2-} ions makes the soil acidic, which renders it unfit for plant growth. The acidic soil can be reclaimed or improved by liming, i.e. by addition of $CaCO_3$.

$$\text{Soil]}\ H^+ + CaCO_3 \rightleftharpoons \text{Soil]}\ Ca^{2+} + CO_2 + H_2O.$$

In highly acidic soils, the bacterial population decrease and the population of fungi increases. Pathogenic fungi, such as plasmodiosphora brasicate are formed in acidic habitats. Also there is excessive growth of weeds and herbs.

In case, sodium or potassium nitrate are used as fertilizers again and again or in excessive amounts, the NO^-_3 ions are used up by successive crops, but the Na^+ and K^+ ions get accumulated in the soil. These cations render the soil alkaline thereby making it unfit for subsequent crop cultivation. Alkaline soils can be reclaimed or improved by treating the soil with $Al_2(SO_4)_3$ or $Fe_2(SO_4)_3$ which set free acid on hydrolysis

$$2Fe^{3+} + 3SO_4^{2-} + 6H_2O \longrightarrow 2Fe(OH)_3\downarrow + 6H^+ + 3SO_4^{2-}$$

Alkaline soils can also be improved by the addition of sulphur, which is oxidised in soil by bacterially mediated reactions to H_2SO_4.

$$S + 1.5\,O_2 + H_2O \longrightarrow 2H^+ + SO_4^{2-}$$

Salination of Soils

For a healthy crop yield, it is necessary to use fertilizer and the fields also need to be well irrigated. In case there is no outlet for unused water, the soil becomes highly saline (salt containing). This is because the salt formed gets accumulated in the soil. The salime soil becomes unfit for growing crops. Reclamation of saline soil requires removal of sodium and calcium. It can be achieved by the application of gypsum followed by leaching. The sodium sulphate formed can be leached out by irrigation.

Due to excessive use of nitrogen fertilizers, concentration of nitrate ions increases in leafy vegetables. Consumption of such leafy vegetables, results in bio accumulation of nitrates in the bodies. Nitrate itself is not harmful but in human body, it is reduced to nitrite, which is a probable factor for the occurrence of cancer. Excessive amounts of nitrates may also lead to **blue baby syndrome** or **methaemoglobinaemia** in small children. In the body, as already stated, the nitrates gets reduced to nitrites, which react with haemoglobin (in the blood stream) and reduce the oxygen carrying capacity of the blood. This condition is referred to as **blue baby syndrome**. The symptoms include diarrhoea and vomitting and the child's complexion become stale blue, which may be fatal.

Another major problem associated with the use of excessive amount of fertilizers is **eutrophication.**

The problems associated with the use of excessive amounts of fertilizers can be overcome by the following:

- Rotation of crops on scientific basis.
- The amount of fertilizers should be used in accordance with professional advice. Only the recommended dose should be used. It is safer to use less amount than required.
- It is best to use organic manure as fertilizer.

Pesticides

Pesticides are chemical substances that are used to kill or block the reproductive process of unwanted or harmful organisms (pests), which are responsible for damaging the crops. Pesticides, in a way, are essential for increasing the yields of crops. However, according to a study conducted by WHO about 7,50,000 people are poisoned by pesticides every year, resulting in about 14,000 deaths. The term, pesticides includes insecticides, fungicides and herbicides.

Insecticides

These are used to eliminate the insects, which eat away the crops. Besides this, the insecticides are also responsible for stopping the spoilage of food crops (e.g. grains) during storage by controlling the insects. They are also useful for controlling the spread of diseases like sleeping sickness, bubonic plague and malaria. They also control household insects like cockroaches, houseflies and mosquitoes. The synthetic insecticides, broadly speaking, are of the following four types.

(*a*) **Organochlorines.** These have been in use since 1950. One of the most common and best known insecticide is DDT [2, 2-bis (p-chlorophenyl)-1, 1, 1-trichloroethane]. Other examples include aldrin and dieldron.

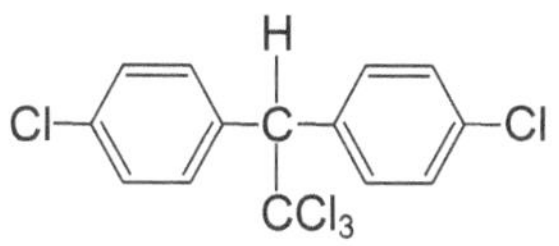

DDT

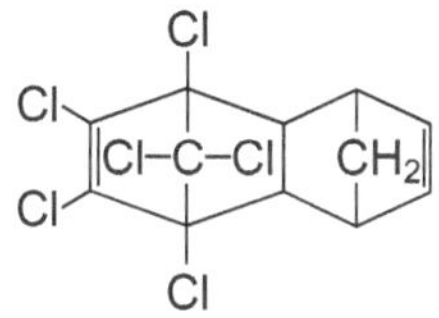

Aldrin

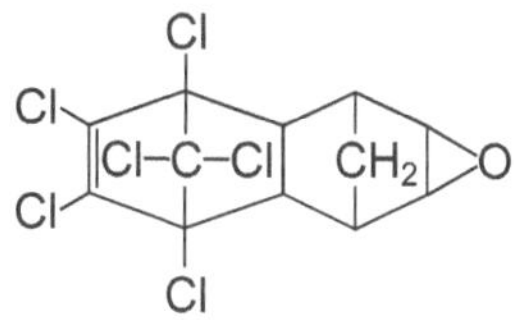

Dieldrin

These insecticides kill not only the unwanted insects (responsible for destroying the crops) but also destroy non-target organisms (which are helpful for the crops. Also these insecticides being non-biodegradable keep on accumulating in the environment and are responsible for pollution. They also get accumulated in the vegetables, fruits and even water. These are harmful for humans beings and have been phased out. Also by continued use of these insecticides, organisms become resistant to them.

(*b*) **Organophosphates.** These insecticides have very low persistence in the atmosphere and are biodegradable. Being toxic in nature, the organophosphates kill not only the non-target insecticides, but are also harmful for humans and wildlife. They are unstable and so have to be applied at frequent intervals. Some examples include parathion, diazinon, bis-(diethylphosphoric) anhydride and demeton.

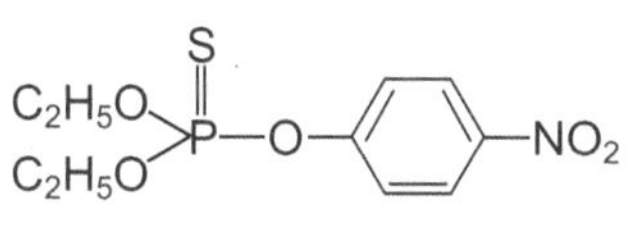

Parathion

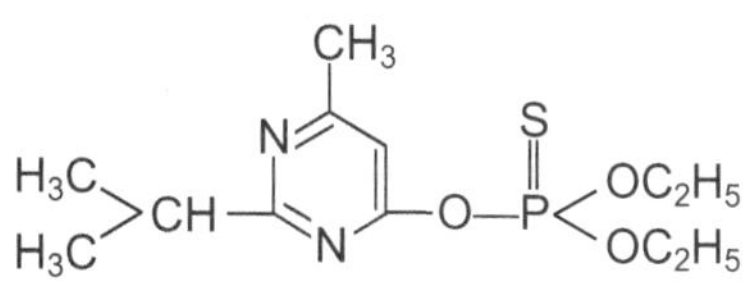

Diazinon

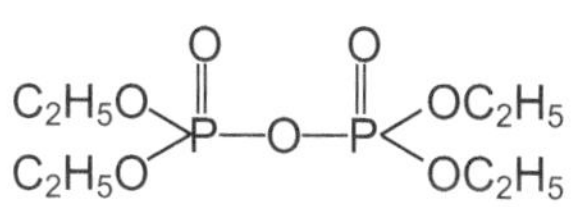

bis (diethylphosphoric) anhydride

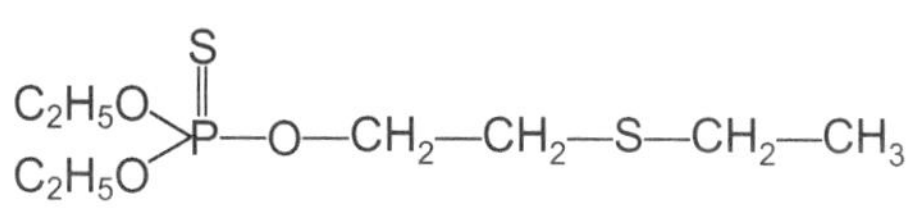

Demeton

(*c*) **Carbamates.** These insecticides are less toxic than organophosphates and less resistance than organophosphates. Some examples include carbaryl and methiocarb

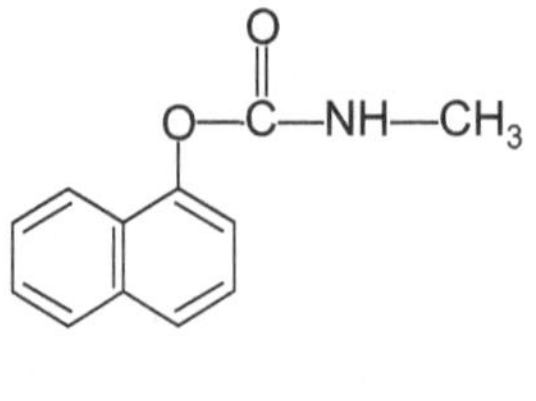

Carbaryl

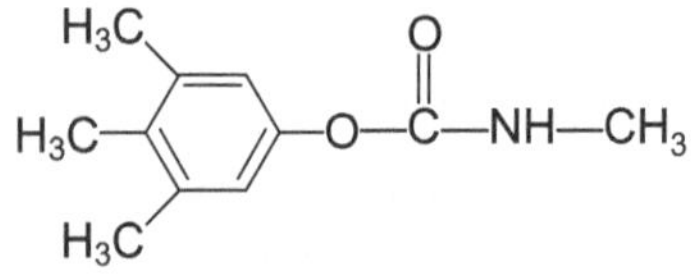

Methiocarb

(*d*) **Pyrethroids.** These are the recent group of insecticides and are the structural analogues of pyrethrum, the naturally occurring insect repellent present in chrysenthesmum. These are neither persistant nor toxic, though they are costly. About 33 per cent of the worlds insecticides are pyrethroids. Some examples are pyrethrum and deltamethrin.

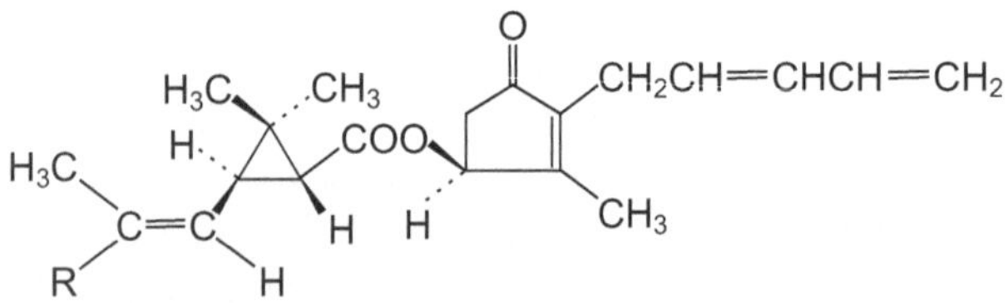

Pyrethrum

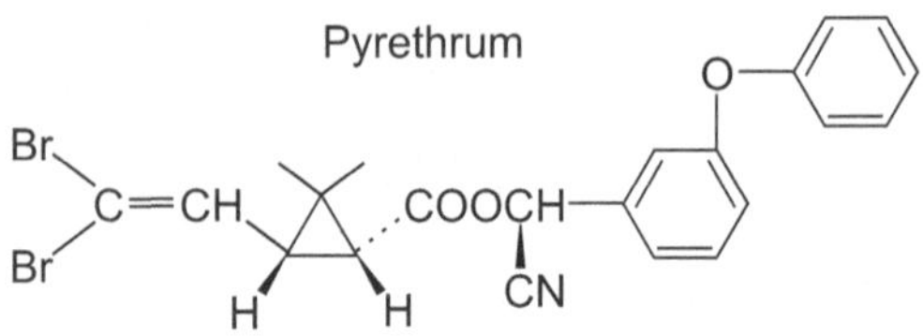

Deltamethrin

Effect of Insecticides on Environment: Excessive use of insecticides affect the environment in several ways. Some of them are:

- Insecticides accumulation in the fatty tissues of various organisms including man via plants and vegetables, etc. Accumulation in food chain has disastrous effects.
- Insects develop resistance against insecticides after a certain period of time. Thus, the effectiveness of the insecticides decreases considerably.
- Accumulation of insecticides (e.g. DDT) in the fatty tissues of birds considerably decline their capacity to reproduce resulting in the decline of their population.
- Contamination of river and ocean water with insecticides effects fish and other marin creatures.

- The workers in industries manufacturing insecticides are affected by slow poisoning effects of the insecticides.
- Accidents in manufacturing units lead to environmental pollution. The worlds worst accident happened in Bhopal (Madhya Pradesh, India) on 2 December 1984, when mehyl isocyanate gas (MIC) used in the manufacture of insecticide got leaked from storage tanks. This accident was responsible for the death of more than 20,000 people and disability to lakhs of people (see Chapter 6).

Since insecticides are essential for agricultural production, it is not possible to ban the manufacture of such inscticides which cause environmental problems. The best option is to look for alternative insecticides, which only kill the pests but are not harmful to the environment. Following are two types of insecticides, which have been used to control insects.

(*i*) **Hormonal Insecticides.** These insecticides kill the insects by upsetting their normal metabolic process. A hormone (Juvenile hormone) shows insecticidal properties. This hormone is initially present in insect larva but its production stops before undergoing metamophosis (i.e. when it grows into adult insects). In case the larva is continuously exposed to the Juvenile harmone, the larva cannot mature—the result is that its production stops.

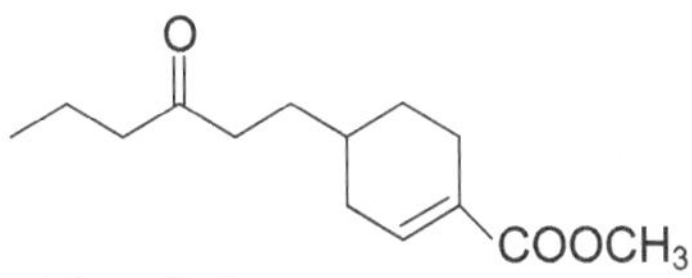

A juvenile hormone

(*ii*) **Sex Attractants.** A group of compounds, known as pheromones have the capacity to attract the male insects from a long distance. If these sex attractants in exceedingly low concentration are kept at a place, then male insects are attracted towards it from a radius of about 3 km. Once assembled in a small area, they can be killed by any means.

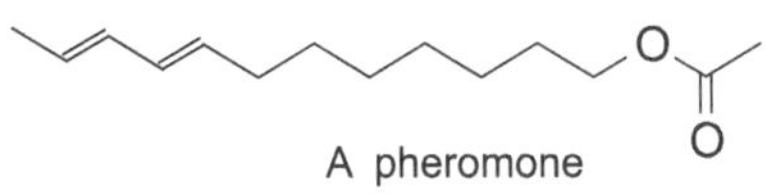

A pheromone

Fungicides

These are used to check the growth of fungi. Fungi are plants without chlorophyll. So they cannot use solar energy to convert CO_2 and H_2O into carbohydrates (photosynthesis). They live as saprophytes on decaying organic matter or as parasites at the expense of living organisms. Fungicides play a vital role in controlling the growth of fungi. Broadly speaking, fungicides are of two types.

(*i*) **Systemic Fungicides.** These fungicides, when applied to the plant, rapidly distribute itself in various plant tissues. Some examples are:

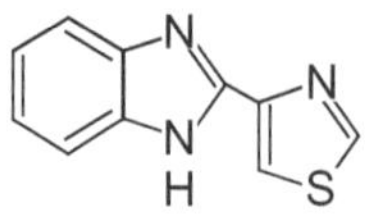

Thiabendazole

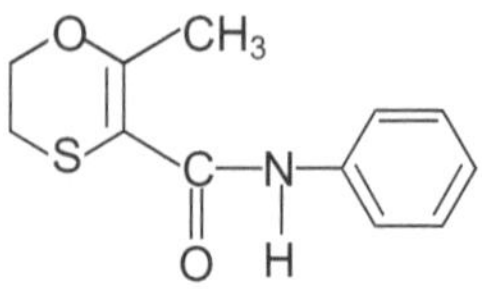

Carboxin

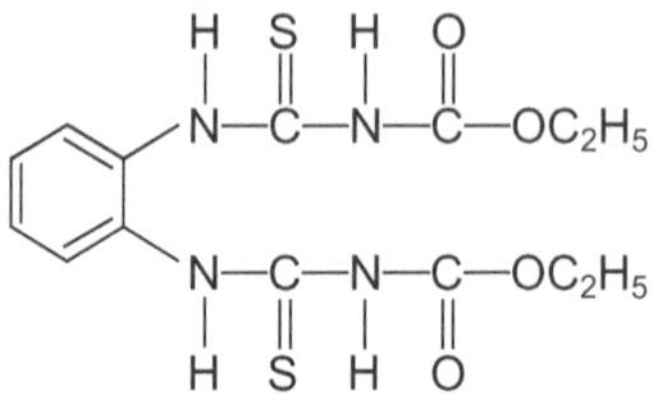

Thiophanate

(*ii*) **Non-systemic Fungicides.** These protect only that parts of the plant where it is applied. Some examples are:

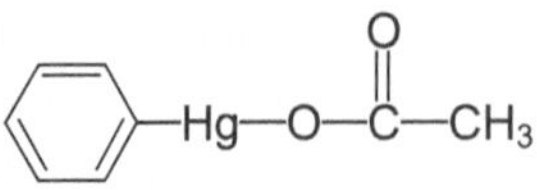

Phenylmercury acetate

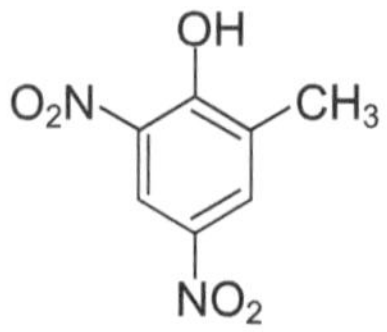

4,6–Dinitro–o–cresol

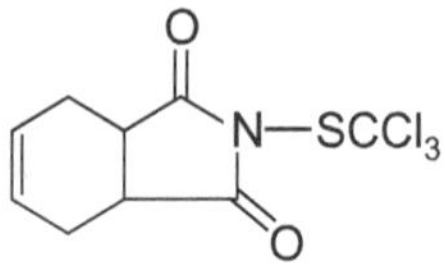

Captan

Effect of Fungicides on Environment

Fungicides have the following effect on the environment.

- In the soil when fungi are eliminated by the use of fungicides, the bacterial population increases. In this way the plants become susceptible to bacterial diseases.
- Certain fungicides are toxic to soil arthropods. For example, captan at a higher dose reduce the population of the useful soil invertbrates like springtails, collembola, earthworm and millipedes.
- Mercural fungicides, e.g. phenyle mercury cyanide used in seed treatment cause destruction of the seed eating birds like bobwhite quaal and pigeons. These insecticides also impair the egg hatching in pheasants, resulting in reduction of population of such species.
- Certain fungicides e.g. captan inhibit the growth of fresh water algae even in very low concentrations.
- Continued use of pesticides makes the pests (that are desired to be eliminated) resistant to them.
- Mercural fungicides are also responsible for human poisoning and deaths. This happens when ingestion of flour and wheat seeds treated with organo-mercury fungicides leads to mercury poisoning. Such poisonings have been recorded in Iraq in 1956, 1960 and 1972, in guatemala in 1965 and in Pakistan 1969.

Redressal of the Effects of Fungicides

The environmental effects of fungicides can be minimised by using them selectively in low concentrations. Attempts should be made to develop new fungicides, which should be more effective, less toxic and more environmental friendly. One such fungicide is a derivative of β-methoxyacrylic acid.

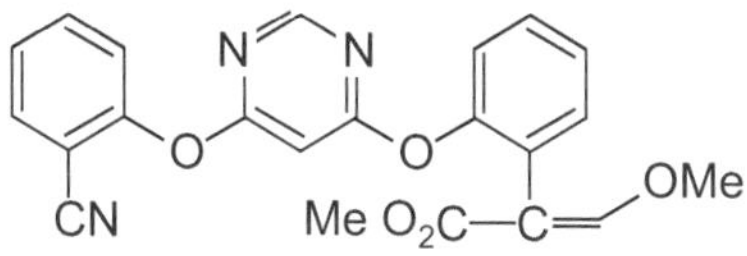

It is likely to be a very useful anti-fungal agent, since it is effective against a wide range of fungal pathogens.

Herbicides

Herbicides are chemicals that kill unwanted weeds or interfere with their growth. Weeds are unwanted plants which grow at places where they are not required, e.g. in agricultural crops and deplete the nutrients which are meant for the betterment of the crops. The weeds retard the growth of the main crops. Herbicides are of two types:

(*i*) **Contact Herbicides.** These herbicides kill the weeds by direct contact and are quick acting. Some examples include

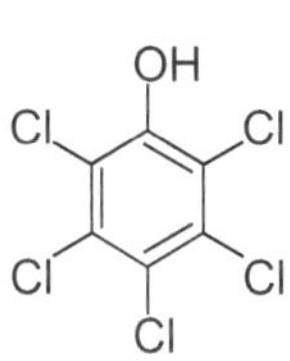
Pentachlorophenol

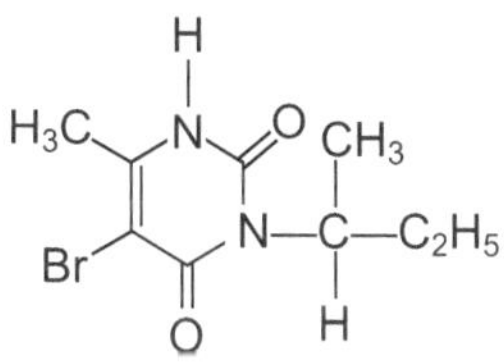
Bromacil

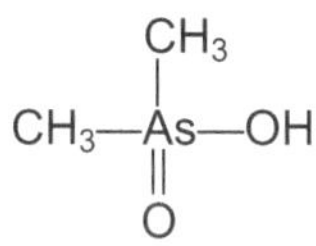
Cacodylic acid

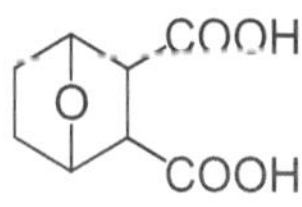
Endothal

Contact herbicides are ineffective against those weeds whose roots produce new shoots at a fast rate.

(*ii*) **Systemic Herbicides.** These diffuse into the plant's body and are very useful. Examples include the following.

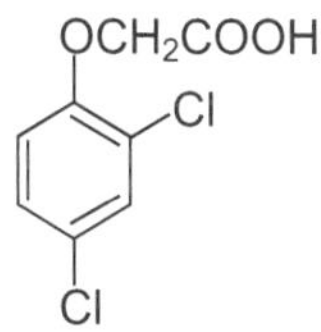
2, 4-D

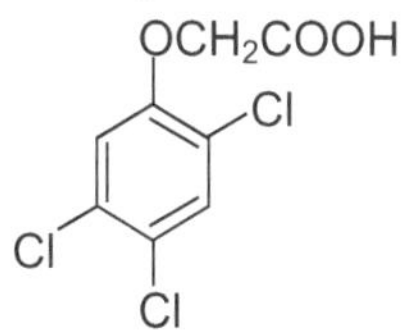
2, 4, 5-T

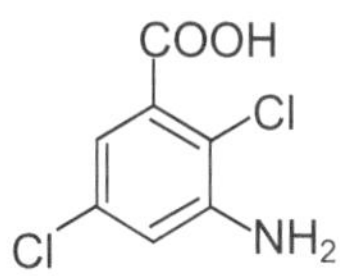
Chloramben

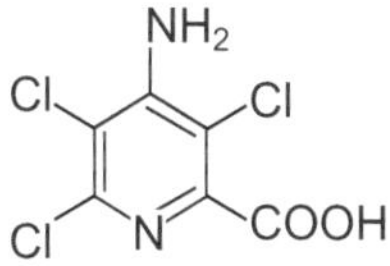
Picloram

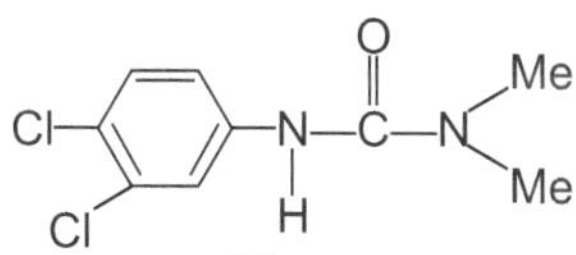
Diuron

Systemic herbicides are effective at low concentrations.

Herbicides act as growth inhibitors for the weeds. For example, the contact herbicide, pentachlorophenol acts as mitotic poison. Some herbicides (e.g. diuron) act as a photosynthetic inhibitor. Finally some herbicides (e.g. picloram) act as a respiratory inhibitor.

Effects of Herbicides on Environment

Like fungicides, herbicides also are extremely important for agricultural production. However, these have the following effects on environment.

- Residues of herbicides in soil increase fungi count resulting in fungal diseases.
- Hyerbicides like 2, 4-D and 2, 4, 5-T when sprayed of the trees, etc., are harmful to millards, quails and partridges.
- Herbicides are hazardous to live stock which eat weeds exposed to sub lethal doses of 2, 4-D and 2, 4, 5-T.
- The production of fresh water algae is reduced due to 2, 4-D and 2, 4, 5-T.
- Herbicides are extremely harmful to the health and survival of human race. On being exposed to 2, 4-D and 2, 4, 5-T which are used to defoliate forest areas (as in the case of Vietnam War) produce birth defects, e.g. children being born with deformed or missing limbs, etc. This is attributed to damage of male sperm cells.

The weeds develop resistance to the conventional herbicides by their continued application. The best way to solve this problem is to develop new and novel herbicides. One such herbicides is δ-aminolevulinic acid. It uses solar energy to produce toxic components that kill the weeds. On exposure to sun light δ-aminolevulinic acid is converted into magnesium tetrapyrrole, which is precursor to chlorophyll synthesis.

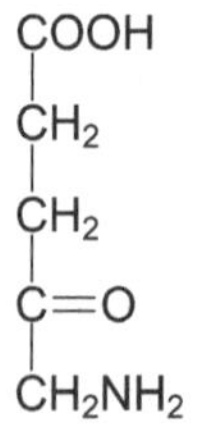

δ-Aminolevulinic acid

The magnesium tetrapyrrole absorbs light energy and undergoes photochemical reaction producing singlet oxygen, a highly reactive form of oxygen. The liberated free radical kills the weeds. The left over δ-aminolevulinic acid or magnesium tetrapyrrole is biodegradable and so does not cause any problems.

Plastics and Polymers

A polymer is a macromolecule, having very high molecular mass and is formed by joining of a large number of repeating units called monomer by the process called **polymerization**. A plastic is also a polymer (but all polymers are not plastics).

It can be moulded into various shapes. The term 'plastic' is used to include many synthetic polymeric materials like rubbers, fabrics, resins, etc. Polymers (including plastics) form the backbone of modern civilisation. In fact, in all walks of life, starting from household utensils, cloths, furniture, etc. to automobiles to spacecrafts, polymers and plastics are extensively used.

Plastic are of two types—Thermoplastics and thermosetting plastics. The thermosetting plastics undergo reversible change when heated and cooled. On heating, they soften and flow; on cooling they harden. Examples of thermoplastic include polyethylene, teflon, nylon 66, polyvinyl chloride and polystyrene.

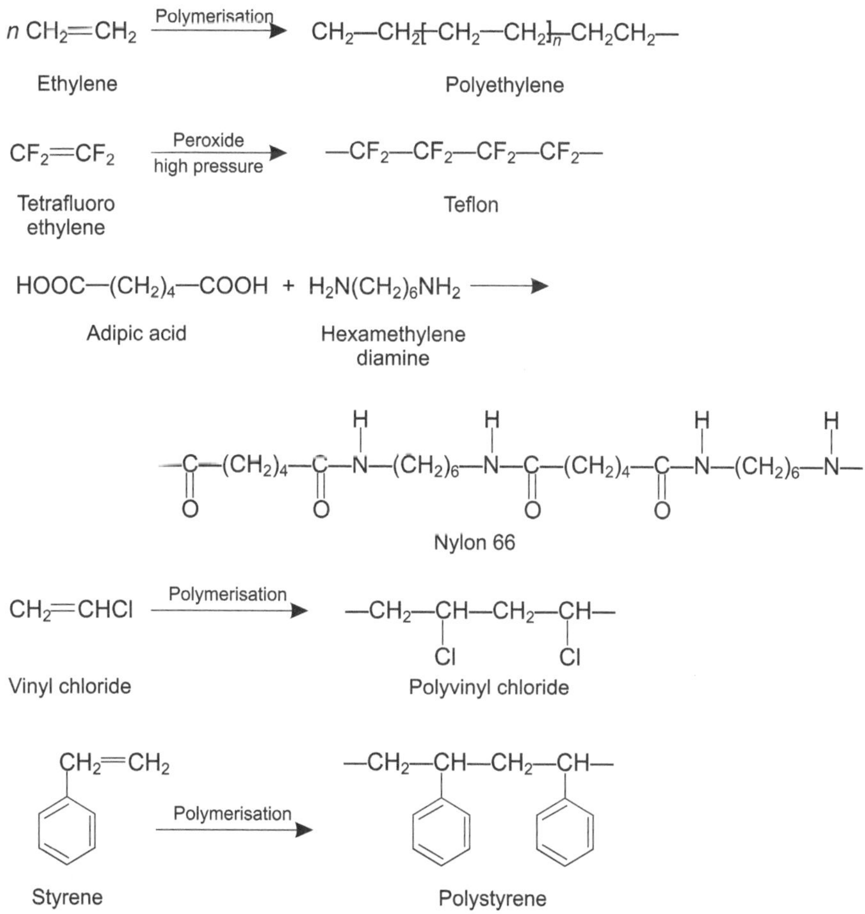

Thermosetting plastics, on the other hand, are those which may be heated to a fluid state only once. The chemical structure of thermosetting plastic is altered by heat, and cross-linked products are formed, which cannot be resoftened. Examples include bakelite, melamine-formaldehyde and polyurethane.

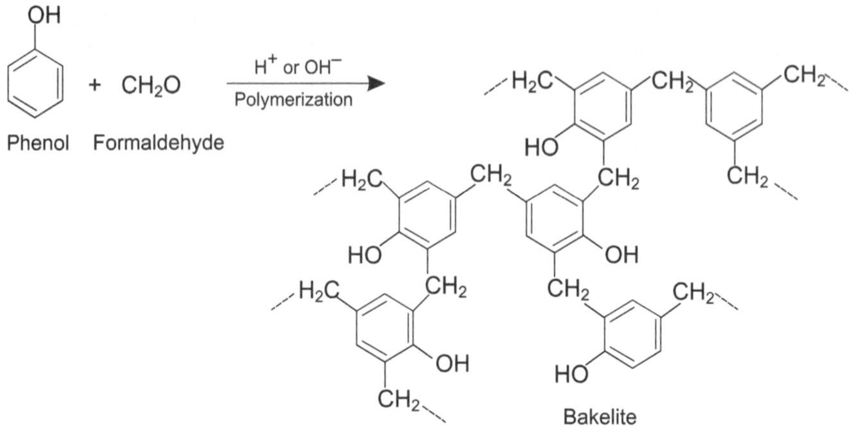

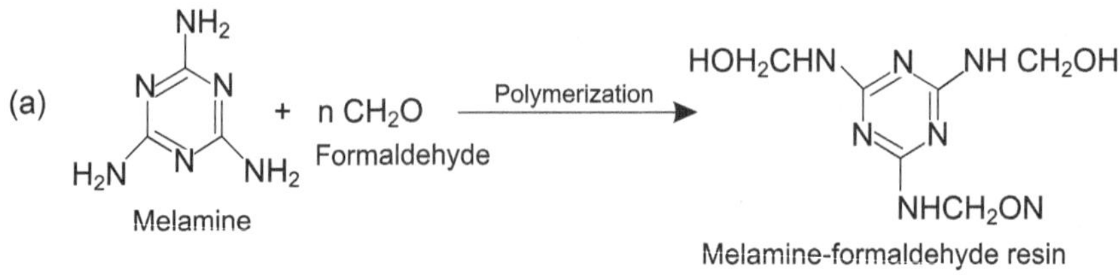

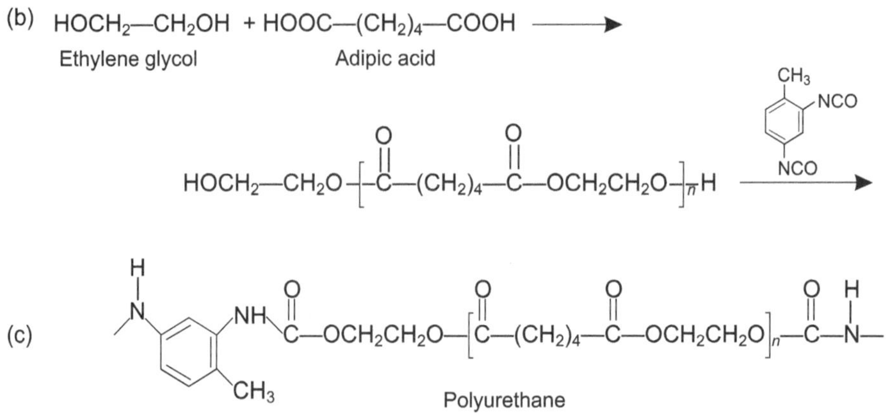

It is a common knowledge that polymers including plastics are non-biodegradable. However, biopolymers like starch, cellulose, proteins and nucleic acids are biodegradable. A major portion of synthetic polymers including plastics is used as throw away containers and packing materials; these do not disintegrate by themselves, i.e. they are not biodegradable and they do not degrades over a long period of time. This property makes the polymers and plastics stay intact for a much large periods—an advantage for use in household items, etc. This non-biodegrability of polymers and plastics has caused a serious waste disposal problem for the mankind. With the ever increasing use of plastics, the time is not far away when the entire civilisation will be buried under a pile of plastic debris. The waste plastic causes acute environmental problems in all the segments of the environment, i.e. the earth, the water bodies and the air. In view of the above efforts have been and are being made by researchers all over the world to evolve such plastics which should be biodegradable.

An important reason for polymers and plastics being non-biodegradable under natural conditions is their hydrophobicity (water repelling capacity). In case the hydrophobic character is improved, the microorganisms will be able to degrade the plastics, though this would mean shorter span of life for them. It is best to achieve a balance between the disadvantage arising out the durability of plastics and the disadvantage arising out of their non-biodegradability in environment. The objective could be achieved by adding some additives, which after been introduced in the final product, make these biodegradable. For example, plasticisers like di-(2-ethylhexyl) phthalates, di-(2-ethylhexyl) adipate and polychlorinated biphenyl's are added to make plastics more flexible. These plasticisers get incorporated between the polymer chain, weakening the forces of attraction between them and so they increase their flexibility.

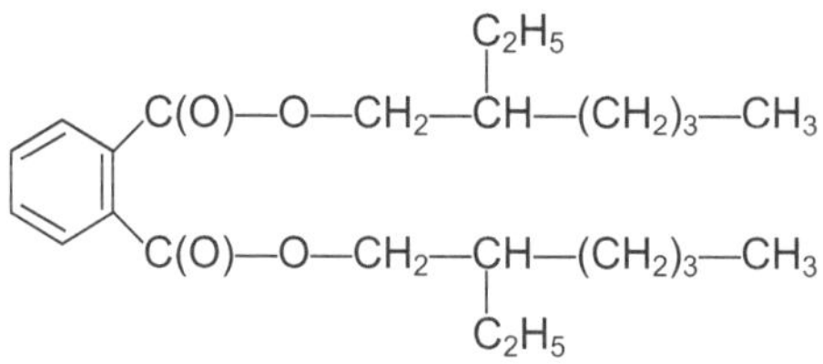

Di-(2-ethylhexyl)phthalate

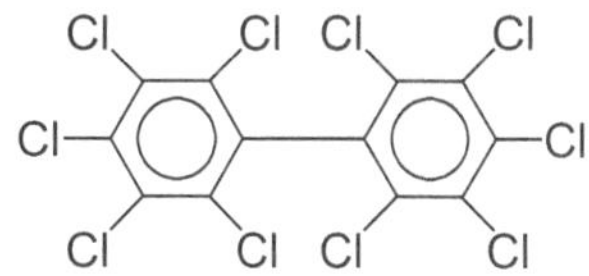

Polychloriated biphenyl

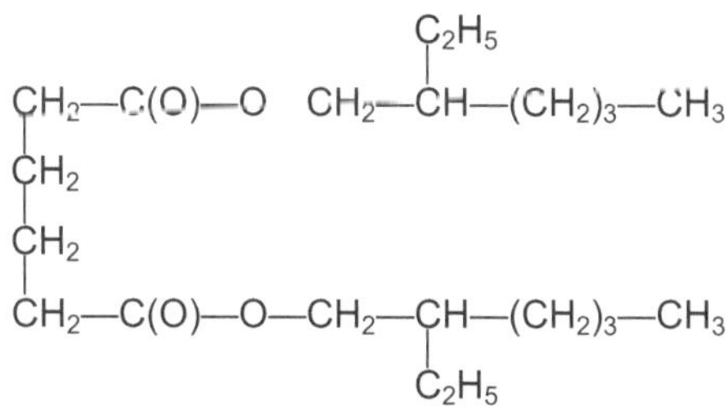

Di-(2-ethylhexyl) adipate

Effect of Polymers and Plastics on Environment

- 2-chloro-1, 3-butadiene (chloroprene) used for the production of synthetic rubber (neoprene) can cause liver damage.
- Toluene -2, 4- diisocyanate is responsible for acute respiratory problems.
- Vinyl chloride used in the manufacture of polyvinyl polymer results in rare form of liver cancer.
- Polychlorinated biphenyls are released in the atmosphere by burning waste plastics in incinerators. These chemicals are washed down into the aquatic systems. Being fat soluble, they get accumulated into aquatic organisms.

When human beings consume fish, they suffer from an ailment called yusho disease; the symptoms are headache, fatigue, pain in joints, anaemia. Concentrations above 12 ppm can cause blindness.

- The polymers and plastic waste materials dumped on the land cause major problems of their disposal. The non-biodegradable nature of these materials makes them a permanent waste. In case, the waste material is incinerated, the problems associated with air pollution arise. The gases evolved by are CO_2 (responsible for green house effect), oxides of nitrogen (responsible for acid rain) and carbon monoxide, an extremely poisonous gas. Incineration also releases particulate matter into the atmosphere. In case asbestos has been used as a filter during the manufacture of plastics, an incineration of plastics, asbestos is injected in the air which is responsible for several health hazards.

Checking Polymers and Plastic Pollutants

To a certain extent, some of the polymers can be recycled. This technology has been successfully used for polyvinyl chloride containers and also for recycling of polymers, e.g. polyester, polyethylene, terephthalate, etc.

Some of the polymers can be converted into monomers, which can be converted back into polymers.

The best option is to make polymers which are biodegradable.

Chemical and Metallic Pollutants

Metals are one of the pollutants which can neither be degraded nor metabolised and so these remains in the environment for an extremely long time. These metals pollute all the components of the environment, viz., atmosphere, land and aquatic systems.

The chemical and metallic pollutants result from a large number of industries including textiles, pesticides, dyes, soap, synthetic detergents, tanneries, drugs, batteries, cement, asbestos, rubber, petroleum, paper and pulp, glass, steel, electroplating and metal extraction processing industries.

A number of toxic metals such as arsenic, cadmium, chromium, copper, lead, mercury and selenium are added to the soil in one form of the other. Synthetic chemicals and fertilizers are a source of trace metals which are added to the soil. Sources of some metallic pollutants and their effects are discussed in (Chapter 2).

Biological Agents

Soil and crops are contaminated by biological agents. These biological agents occur naturally in contaminated soil or are discharged into the soil by excreta of birds, animals and humans. Also, municipal wastes including domestic wastes and garbage are responsible for the discharging of biological agents into the soil.

Acid Rain

Air contains pollutants like SO_2 and NO_2 (see Chapter 4). SO_2 present in the air undergoes catalytic and photolytic oxidation to form SO_3, which reacts with moisture present in the atmosphere to form H_2SO_4.

$$SO_2 + O_2 \longrightarrow SO_3$$

$$SO_3 + H_2O \longrightarrow H_2SO_4$$

Also, NO_2 present in the atmosphere reacts with rain water or moisture in presence of O_2 or O_3 and produce HNO_3.

$$4NO_2 + 2H_2O + O_2 \longrightarrow 4HNO_3$$

$$2NO_2 + H_2O + O_3 \longrightarrow 2HNO_3 + O_2$$

H_2SO_4 and HNO_3 formed (as stated above) come down the atmosphere in the form of acid rain. This damages the crops and plants. Also the soil becomes acidic, which is harmful for subsequent crops (see Chapter 4).

3.4 WASTE TREATMENT

Waste product is defined as any product of metabolism that is not required for further metabolic processes and is therefore excreted. Common waste products also include nitrogenous compounds such as amnonia, carbondioxide and also radioactive wastes, municipal wastes which include domestic house hold wastes. A major amount of wastes are generated from various industries. There have been several practices for managing such wastes. Some other are:

- Produce less waste
- Convert hazardous substances into less hazardous ones
- Proper and safe disposal of waste.

3.4.1 Reduction in the Amount of Waste

This is the safest and least expensive procedure. The amount of hazardous waste generated by manufacturing and other processes can be reduced by changes in the chemical processes involved, equipment and raw materials used. It is best to follow the principals of green chemistry. This aspect forms the subject matter of Chapter 6.

Recycling and reuse also eliminates hazardous wastes. For example, certain solvents can be recovered and reused or recycled. This practice yields economic benefits both in terms of reducing the cost and the handling of hazardous waste. The waste generated from metallurgical processes can be recycled. Also the scrap metals could be recycled.

Certain industries can mutually benefit by taking wastes from one industry and using it by another industry.

3.4.2 Conversion to Less Hazardous Substances

The toxicity of hazardous characteristics can be reduced by physical, chemical or biological methods.

Physical methods such as absorption on charcoal or resin filters for absorbing toxins can be used. Distillation is also useful at times. Precipitation and immobilisation are also helpful in making wastes suitable for long term storage.

Chemical methods are useful for detoxifying the wastes. These include neutralisation, oxidation, reduction, ion exchange, etc.

Biochemical methods involve the use of bacteria, soil microorganisms and other species in reducing the hazardous characteristic of the waste.

Biodegradation can be accompanied by aerobic and anaerobic methods.

Incineration is a physical method of treatment. It involves heating of the waste to about 1200°C for quite some time to complete destruction. In this procedure the volume of waste is reduced by at least 90 per cent and the resulting waste is safar to store in a land fill or other storage site. The products of combustion are CO_2, H_2O and other gases. It is a quick, easy and permanent solution for many hazardous wastes.

3.4.3 Disposal Methods

Land filling has been a very common and popular method for the disposal of hazardous waste. However, the land fills must be secure, so that the waste material is not leached into ground water. For this purpose, the land fills have a bottom layer of gravel and plastic lining.

Alternatively, the wastes are put into deep wells. The waste is injected into a permeable rock layer several thousand meters below the surface, in geological basins. This method requires careful monitoring after disposal.

Earlier, ocean dumping for disposal of wastes was used. This method has been discontinued due to the consequences of marine pollution.

EXERCISE

1. How is soil formed?
2. Describe the different horizons of soil.
3. Give an account of macro and micro nutrients present in the soil.
4. Discuss the various sources of pollution of the soil.
5. How do fertilizers affect the soil?
6. How are acidic, basic and saline soils reclaimed?
7. Discuss the affect of insecticides on the environment.
8. Write notes on fungicides and herbicides.
9. Discuss the effect of soil pollution by plastics and polymers and metals.
10. How is acid rain caused?
11. Write a note on waste treatment.

Chapter 4

Atmosphere

4.1 INTRODUCTION

The gaseous mixture surrounding the earth is called the atmosphere. The composition of the atmosphere at sea level is nitrogen (78%), oxygen (21%), argon (0.9%) carbon dioxide (0.3%) along with inert gases in trace amounts. Besides these, the atmosphere also contain varying amounts of water vapour and oxides of nitrogen and sulphur. The atmosphere becomes thinner as one goes up from sea level. The temperature, pressure and composition of the atmosphere also changes with height. About three fourth of the mass of the atmosphere is within 11 km from the surface of the earth. The atmosphere is divided into four regions. These are the troposphere, stratosphere, mesosphere and thermosphere.

The life on earth is protected by the atmosphere by absorbing the harmful ultraviolet radiations from the sun.

4.2 TEMPERATURE AND THE ATMOSPHERIC LAYER

As already stated, the temperature of the atmosphere varies with altitude and that the atmosphere is divided into following four layers or region.

Troposphere. The region nearest to the earth and extending up to an altitude of 10 km is called troposphere. In this region, the temperature decreases with increase in height from about 15°C (at ground level) to about –56°C. This decrease in temperature is attributed to fall in the pressure and density of air. About 75 per cent of the mass of the atmosphere is present in this region. In the absence of any significant air pollution, the composition of air in atmosphere is more or less constant.

Stratosphere. This region is above troposphere and extends up to a height of 50 km. As the height increases, the temperature increases from –56°C to –2°C. The increase in temperature is attributed to the absorption of ultraviolet radiation from the sun. In this region ozone is produced from oxygen by sun rays.

$$3O_2 \xrightarrow{\text{Sun rays}} 2O_3$$

The ozone forms a stable layer in the stratosphere. This layer protects life on earth by absorbing the harmful UV radiations from the sun.

Mesosphere. The region above stratosphere and extending up to a height of 90 km is known as mesosphere. In this region, the temperature decreases from –2° C to – 92° C with increase in height. The decrease in temperature in this region is due to comparatively low levels of ozone present.

Thermosphere. The region above mesosphere and extending up to 500 km is known as thermosphere. In this region the temperature rises from –92° to 1200°C. Oxygen and hydrogen present in this region absorb solar radiation and undergo ionisation.

The region of the atmosphere between its altitude of 50 km to 100 km is called **Ionosphere**. In this region, appreciable levels of ions like O_2^+, O^+, NO^+ and electrons are present. These are formed due to solar radiation.

Following figure depicts a simple way of representing various regions of the atmosphere.

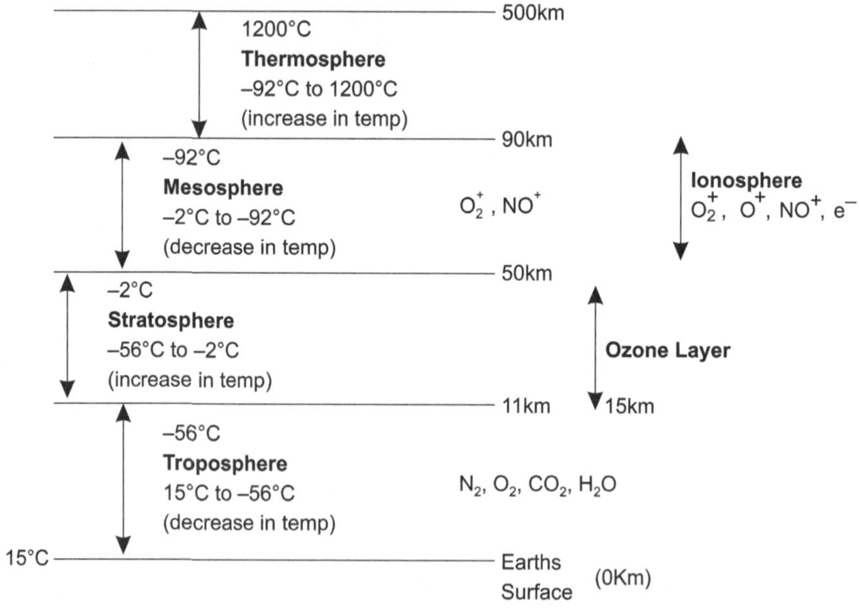

Fig. 4.1 Simplified way of representation of various region of the atmosphere.

The air present in the atmosphere consists of a number of gases. The composition of the earth's atmosphere is as follows:

- 75.523% nitrogen
- 23.133% oxygen
- 1.288% argon
- 0.053% carbon dioxide
- 0.001267% neon
- 0.00029% methane
- 0.000033% krypton
- 0.000724% helium
- 0.0000038% hydrogen

4.3 PARTICLES, IONS AND RADICALS IN THE ATMOSPHERE AND THEIR FORMATION

Particles are present in the troposphere. They vary in number from several hundred per cubic centimeter in pure air to more than 10^5 per cubic centimeter in polluted air. The particles also vary in size (0.1 – 10μ). Their mass ranges from 10μg/m^3 in clean air to 60 – 2000 μg/m^3 in polluted air.

Particles of colloidal size in the atmosphere are known as aerosols. **Aitken particles** are also aerosols of natural origin having diameter less than 0.2μ. Such particles originate from vegetation. Bacteria, fog, pollen grains and volcanic ash are other particles of natural origin in the atmosphere. In case of particles, their size is more important than their total count. Particles in the size 0.1–1μ are responsible for the electrical phenomena in the atmosphere. They are also responsible for cloud and fog formation. Besides, particles serve as nucleic for the formation of ice—crystals and water droplets. The particles of metals also have catalytic effects on oxidation and photochemical oxidations. Finally, particles play an important role in determining the heat balance of the earths atmosphere via light reflection.

4.3.1 Formation of Particulate Matter

Inorganic particulate matter originates from the following sources:

- Automobile exhaust discharge lead particles is the atmosphere. It is estimated that in USA more than 2×10^5 tonnes of lead are dumped in the atmosphere every year. However, with switching off to unleaded petrol, this quantity is decreasing over the years.
- Combustion of pyrite containing coal dumps besides carbon particles in the form of ash also metal oxides.

$$3FeS_2 + 8O_2 \longrightarrow Fe_3O_4 + 6SO_2$$

The formation of aerosols is explained by the atmosphere oxidation of SO_2 to SO_3 followed by reaction with moisture to produce H_2SO_4. The acid produced reacts with basic air pollutants like NH_3 or CaO to form salts

$$SO_2 + O \longrightarrow SO_3$$
$$SO_3 + H_2O \longrightarrow H_2SO_4$$
$$H_2SO_4 + 2NH_3 \longrightarrow (NH_4)_2SO_4$$
$$H_2SO_4 + CaO \longrightarrow CaSO_4 + H_2O$$

In a similar way the oxides of nitrogen react with moisture to produce HNO_3. The acid produced reacts with basic air pollutants to produce nitrates.

$$NO + O \longrightarrow NO_2$$
$$4NO_2 + 2H_2O + O_2 \longrightarrow 4HNO_3$$
$$HNO_3 + NH_3 \longrightarrow NH_4NO_3$$
$$2HNO_3 + CaO \longrightarrow Ca(NO_3)_2 + H_2O$$

The net result is the formation of aerosol mist consisting of droplets containing salts of H_2SO_4 and HNO_3.

A typical analysis of particulate matter in the air in united states urban locations revealed the following composition ($\mu g/m^3$), total suspended particulate matter (105), NH_4^+ (1.3), NO_3^- (2.6), SO_4^{2-} (10.6), benzene soluble organics (6.8), Sb (0.01), As (0.02), Cd (0.002), Cr (0.015), Cu (0.09), Fe (1.58), Pb (0.79), Mn (0.10), Ni (0.034), Sn (0.02), Ti (0.04), V (0.05) and Zn (0.07) [Source : Air quality Data 1964 – 65, US Dept: of Health, Education and Welfare, Division of Air Pollution, Cincinnati, Ohio 1966 pp. 1 – 2].

Organic particulate matter

is believed to originate from a variety of source like emissions from vegetation and automobiles, combustion of fuels etc. Such particles are in 1μ size range and are potential health hazards. For example, polycylic (polynuclear) aromatic hydrocarbons (PAH) in organic particulate matter have carcinogenic effect. One such compound responsible for carcinogenic effect is beno (α-) pyrene. Such compounds generally occur in urban atmospheres at about 20$\mu g/m^3$ level and originated from the pyrolysis of paraffins present in fuels and plant material.

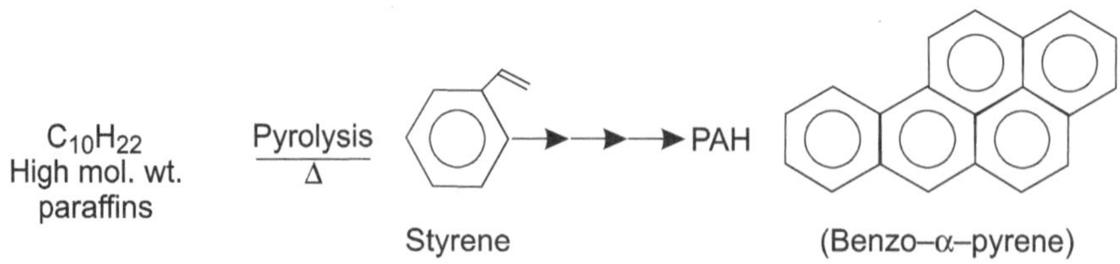

The PAH formed remain adsorbed on soot particles, which are formed as a residue on combustion of fuels in automobiles and electricity generating power plants, which use coal as the raw material. Generally speaking, the soot particles, which we come across contain adsorbed PAH and toxic metals. A typical soot particle is represented is shown in Fig 4.2.

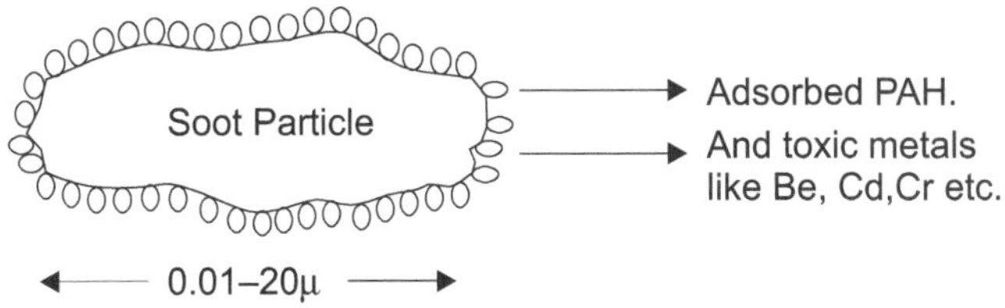

Fig. 4.2 A soot particle contains absorbed PAH and toxic metals

The atmospheric particles (soot particles) enter human body through respiration. These get lodged in the lungs causing health hazards. The removal of particulate matter from the gases is important for controlling air pollution.

4.3.2 Ions and Radicals

Ions are atoms that have either lost one or more electrons, making it positively charged (or cations) or gained one or more electrons making it negatively charged (anions).

The ions are normally present in the ionosphere region of the mesosphere (see p10). In this region, positive ions like O_2^+, O^+, NO^+ etc. and electrons exist at significant levels. These species are formed due to solar radiations. During the night, the UV radiations are not available and so these ions recombine with free electrons to give neutral species from which they originated.

The ions are influenced by the earths magnetic field. This results in the formation of two belts of ionising particles encircling the earth. These are known as van Allen belts.

Radicals are a group of atoms, either in a compound or existing alone. **Free Radicals** are atoms or group of atoms with an unpaired valence electron. Free radicals can be produced by photolysis or pyrolysis in which a bond is broken without forming ions. Because of their unpaired valence electrons, most free radicals are extremely reactive. Besides ions, the atmosphere also contains **free radicals**, which take part in chain reactions in which one of product is a free radical. The chain can be terminated by combination of two free radicals.

$$\underset{\text{Aliphatic Hydrocarbon}}{RH} + O + O_2 \longrightarrow RO\dot{O} + H\dot{O}$$

$$CH_3-CHO + h\nu \longrightarrow \dot{C}H_3 + HC\dot{O}$$

$$\dot{C}H_3 + \dot{C}H_3 \longrightarrow C_2H_6$$

Free radicals play an important role in photochemical smog formation. This aspect forms the subject matter of a subsequent section.

4.4 CHEMICAL AND PHOTOCHEMICAL REACTIONS IN THE ATMOSPHERE

The most important gases which are present in the atmosphere and undergo various types of chemical and photochemical reactions are oxygen, nitrogen and carbon dioxide. Besides there are gases like ozone, nitrous oxide, sulphur dioxide and nitrogen dioxide, which are either produced into the atmosphere or are formed is the atmosphere, (e.g. ozone). Chemical and photochemical reactions of all these form the subject matter of the present discussion.

4.4.1 Oxygen

Oxygen is a colourless ordourless gaseous element, most abundaut in the earth's atmosphere. It is of vital importance for all organisms that carry out aerobic respiration.

Oxygen originates by the process of photosynthesis and plays an important role in troposphere and a key role in stratosphere. In troposphere, oxygen takes part in energy producing reactions like burning of fossil fuels. Thus, methane (in natural gas), reacts with oxygen producing CO_2 and liberating energy.

$$CH_4 + 2O_2 \longrightarrow CO_2 + 2H_2O + 213 \text{ Kcal.}$$

In the stratosphere, oxygen is responsible for the formation or generation of ozone. The oxygen undergoes a photochemical reaction in which energy from the sun decomposes the oxygen molecule into reactive atomic oxygen which in turn combines with molecular oxygen to form ozone.

$$O_2 + h\nu \rightleftharpoons 2O$$

$$O_2 + O \rightleftharpoons O_3$$

The formation of ozone is essential for sustaining life on earth by preventing the harmful UV radiations from reaching the sun.

In the upper atmosphere (viz., mesosphere, thermosphere and ionosphere), ultraviolet radiations cause photochemical dissociation and ionisation of ozone to give rise to oxygen atom, oxygen ion and excited oxygen atoms:

$$O_3 + h\nu \xrightarrow{135-176 \text{ nm}} \underset{\text{Oxygen atoms}}{O + O}$$

$$O_3 + h\nu \xrightarrow{308 \text{ nm}} \underset{\text{Oxygen ion}}{\dot{O} + O_2}$$

$$O + h\nu \longrightarrow \underset{\text{Excited oxygen atoms}}{O^* + e}$$

$$O_2 + h\nu \longrightarrow O_2^+ + e$$

The ultraviolet radiations are not available during night; so these ions recombine with free electrons to give neutral species from which they originated.

4.4.2 Nitrogen

Nitrogen is a colourless gaseous element in the air (about 78 per cent by volume). It is an essential constituent of proteins and nucleic aicds in living organisms.

Nitrogen is essential for the sustenance of life. Though nitrogen by itself is inert and unreactive, its presence in the atmosphere (to the extent of about 78 per cent) is extremely important, since it dilutes or eliminates the harmful effects of pure oxygen (oxygen though supports life, but in pure oxygen most of the organic matter is burned). However, oxygen in about 5–6 per cent concentration is useful for all jobs. The origin of nitrogen in the atmosphere is believed to be from breaking down of nitrates in the soil by denitrification process by some micro organisms

$$4NO_2^- + 2H_2O \longrightarrow 2N_2 + 3O_2 + 4OH^-$$

In the presence of lightening and thunderstorm, the nitrogen of the atmosphere combines with oxygen to form nitric oxide (NO), which in turn combines with oxygen to give nitrogen dioxide (NO_2).

$$N_2 + O_2 \longrightarrow 2NO$$

$$2NO + O_2 \longrightarrow 2NO_2$$

Ozone can also oxidise nitric oxide to nitrogen dioxide

$$NO + O_3 \longrightarrow NO_2 + O_2$$

Nitrogen dioxide reacts with water, producing nitric acid, which is washed down in soil is rain to give acid rain.

$$4NO_2 + 2H_2O + O_2 \longrightarrow 4HNO_3$$

4.4.3 Nitrous Oxide

Nitrous oxide (N_2O) commonly known as laughing gas is not produced by atmospheric oxygen. At present, the atmospheric concentration of nitrous oxide is about 0.31 ppm. The major sources of nitrous oxide are:

- microbial action on nitrogenous fertilizers in soil
- burning of biomas, fossil fuels and forests.

The concentration of nitrous oxide has been increasing at the ratt of 0.2 per cent per year. Nitrous oxide is one of the greenhouse gas and contributes to about 4 per cent in greenhouse effect.

In the stratosphere, nitrous oxide plays a role in the destruction of the ozone layer. The various steps involved in the process and as follows.

$$N_2O + O^{\cdot} \longrightarrow 2NO$$

$$N_2O + O_3 \longrightarrow NO_2^{\cdot} + O_2$$

$$NO_2^{\cdot} + O^{\cdot} \longrightarrow NO + O_2$$

The over all reaction is

$$O_3 + O^{\cdot} \longrightarrow 2O_2$$

4.4.4 Carbon Dioxide

Carbon dioxide is released into the atmosphere by burning of different type of fuels, by the respiration process and by the eruption of volcanoes. The green plants absorbs CO_2 gas from the atmosphere to prepare food (carbohydrates) via the process of **photosynthesis**.

$$6CO_2 + 6H_2O \xrightarrow[\text{Chlorophyll (green plants)}]{\text{Solar energy}} C_6H_{12}O_6 + 6O_2$$

Carbon dioxide is one of the very important green house gases (the others being methane, nitrous oxide, chlorofluorocarbons etc). It does not support life. In fact, animals and human beings kept in CO_2 atmosphere die due to lack of oxygen. Though CO_2 is not considered to be a pollutant but in excess amount, it has adverse effects on the climate (for details see green house effect, section 4.9.2). Before CO_2 was introduced into the atmosphere, the average temperature of earth was –15°C. Thus, without greenhouse gases, the planet earth would be uninhabitable. This aspect has been discussed subsequently.

4.4.5 Ozone

Ozone (trioxygen) is a colourless gas, O_3, soluble in cold water and in alkalis. Liquid ozone is dark blue in colour and is diamagnetic.

Ozone is generated in the atmosphere from oxygen. In stratosphere, oxygen undergoes photochemical reaction in presence of rays of sun when oxygen decomposes into reactive oxygen atoms, which in turn combine with a molecule oxygen to form ozone.

$$O_2 + h\nu \rightleftharpoons 2O$$

$$O_2 + O \rightleftharpoons O_3$$

Depending on the circumstances, ozone can be beneficial or harmful. It is an irritating toxic gas and when breathed produces extremely harmful effects like soar throat, bronchial irritation, coughing and fatigue; much higher concentration can even cause death. Ozone is also lethal to low forms of life including bacteria. In some countries (especially Europe), it is widely used as a disinfectant for drinking water supplies. On the other land, the same ozone gas forms a life saving shield that protects us from the extremely harmful effects of the ultraviolet solar radiation. In the atmosphere, ozone is present as the ozone layer that surrounds the earth (within the stratosphere) centered roughly at some 30 – 35 km from above the earths surface.

In the atmosphere, absorption of ultraviolet radiation by ozone molecule within the ozone layer regeneraties an oxygen atom and a oxygen molecule. The oxygen atom also recombines with oxygen molecule to regenerate ozone.

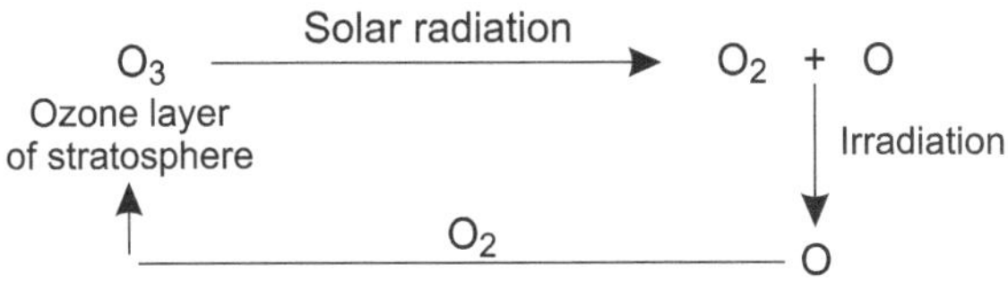

The cycling between the destruction and formation of ozone molecules results in a reasonably constant concentration of ozone that has permitted the life on earth to thrive under the invisible shield. However, if chloro-fluorocarbons (CFCS) are present in thc atmosphere (or discharged into the atmosphere by various routes), the ozone layer is destroyed or depleted causing serious complications.

4.4.6 Sulphur Dioxide

Sulphur dioxide is discharged into the atmosphere as a worst pollutant. Its main sources are eruption of volcanoes, biogenic decomposition of H_2S, burning of fossil fuels, especially coal, petroleum and its products, smelting of sulfide ores, especially those of copper and zinc, manufacture of H_2SO_4 etc. etc.

The sources which emit SO_2 also release H_2S into the atmosphere. H_2S is oxidised to SO_2 by atomic oxygen

$$H_2S + O_2 \longrightarrow HS + OH$$
$$HS + OH + O_2 \longrightarrow SO_2 + H_2O$$

H_2S is also converted into SO_2, by molecular oxygen or by ozone.

$$H_2S + \frac{3}{2}O_2 \longrightarrow SO_2 + H_2O$$
$$H_2S + O_3 \longrightarrow SO_2 + H_2O$$

The sulphur dioxide in the atmosphere is oxidised to SO_3 both by photochemical and non-photochemical processes. SO_2 absorbs strongly in the near ultraviolet region to form electronically excited molecules, which react photochemically with ozone.

$$SO_2 + O_3 \xrightarrow{hv} SO_3 + O_2$$

SO_2 in the excited state also reacts with O_2 in the gas phase to give SO_3.

$$SO_2 + \frac{1}{2}O_2 \xrightarrow{hv} SO_3$$

The above reactions with O_3 and O_2 take place only in presence of uv light, which forms excited molecule of SO_2. In the absence of sun light there is no reaction.

In the non-photochemical oxidation, SO_2 is oxidised by molecular oxygen only in the presence of certain catalysts like oxides of aluminium, calcium, chromium, iron, lead and vanadium. The particulates (e.g. shoot particles) also oxidise SO_2 to SO_3.

$$SO_2 + \frac{1}{2}O_2 \xrightarrow{\text{Particulates}} SO_3$$

The SO_3 formed by any of the above processes, react with water to form H_2SO_4.

$$SO_3 + H_2O \longrightarrow H_2SO_4$$

The formed H_2SO_4 comes to the earth in the form of acid rain. As already stated the H_2SO_4 formed in the atmosphere form aerosols.

4.5 SMOG

Smog is a mixture of smoke and fog. The name smog is derived from combination of smoke and fog.

$$\text{Smoke} + \text{Fog} \longrightarrow \text{Smog}$$

Smog is of two types: classical smog and photochemical smog.

4.5.1 Classical Smog

It is commonly know as smog and is produced from gases like SO_2 and NO_2 present in air (produced by burning of fossil fuels like coal and petrol or gasoline in homes, industries and automobiles). These gases get mixed with moisture in the tiny droplets of fog producing H_2SO_4 and HNO_3 respectively.

$$SO_2 + \frac{1}{2}O_2 \longrightarrow SO_3$$

$$SO_3 + H_2O \longrightarrow H_2SO_4$$

$$2NO_2 + H_2O \longrightarrow HNO_2 + HNO_3$$

$$HNO_2 + \frac{1}{2}O \longrightarrow HNO_3$$

The resulting mixture consisting of H_2SO_4 and HNO_3 get condensed on the solid particulate matter of smoke present in the air to give a colloidal dispersion called smog or classical smog.

Usually smog is formed during winter season and is present in the lower atmosphere. It decreases visibility and affects human health. It causes breathing problems like bronchitis and asthma, which may be responsible for heart problems. It produces irritation to eyes, nose and throat. Due to reduced visibility there are problems in road and rail traffic.

It is well known that in December 1952, a dense cloud it smog was formed over London city and remained for 5 days. It resulted in the death of about 5000 people and created health problems to numerous other people. The cause of death was pneumonia, bronchitis and other respiratory problems. This smog, commonly called London smog formed after the introduction of coal as a fuel, which produced both SO_2 and smoke. The **London smog** is also called Reducing Smog.

4.5.2 Photochemical Smog

This type of Smog is formed by the combination of smoke, dust and fog with air pollutants particularly oxides of nitrogen and hydrocarbons. These pollutants originate from automobile exhaust and other industrial activities.

The formation of photochemical smog can be explained in the following way (chart 4.1). The various steps involved are:

(*i*) The hydrocarbons (which are reactive containing a C=C group) from automobiles interact with ozone to form a hydrocarbon free radical ($RCH_2^{\cdot}$) (1). These free radicals may also be formed by the reaction of hydrocarbons with atomic oxygen [evolved from NO_2 by uv radiation from the sun; NO_2 + uv radiation $\longrightarrow$ NO + O]

(*ii*) The hydrocarbon free radicals (1) react rapidly with oxygen to form another free radical $RCH_2O_2^{\cdot}$(2).

(*iii*) The second free radical (2), in turn, reacts with nitric oxide present in the atmosphere (either from N and O in presence of sun light or by irradiation of NO_2) to produce the third free radical, $RCH_2O^{\cdot}$ (3) (the NO is converted into NO_2).

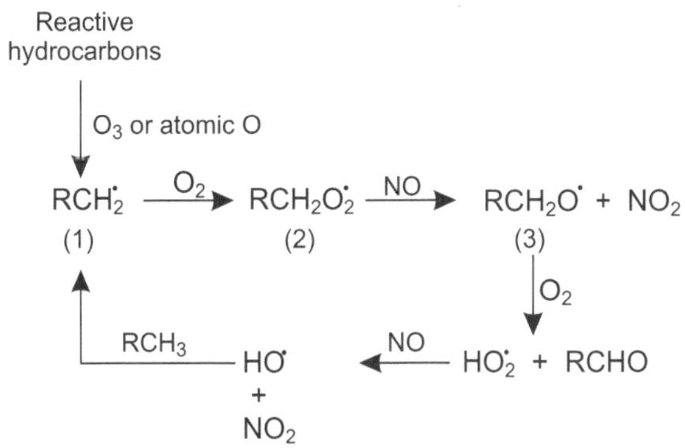

Chart 4.1 Steps involved in the formatter of photochemical smog products

(*iv*) The third free radical, R $CH_2O^{\cdot}$ (3) subsequently reacts with oxygen yielding an aldehyde, RCHO and generating a hydroperoxyl radical ($H_2O^{\cdot}$).

(*v*) The hydroperoxyl radical ($HO_2^{\cdot}$) reacts with nitric oxide to give NO_2 and generating hydroxyl radical ($HO^{\cdot}$).

(*vi*) The hydroxyl radical ($HO^{\cdot}$) being extremely reactive reacts with a hydrocarbon, regenerating the hydrocarbon free radical, $RCH_2^{\cdot}$ (1).

The above cycle is repeated a number of times leading to the rapid build up of photochemical smog products.

(*vii*) The aldehyde (RCHO) formed in step (iv) reacts with hydroxyl radical ($HO\cdot$) producing an acyl radical ($RCO^{\cdot}$), which in turn reacts with oxygen to give peroxyacyl radical ($RCOO_2^{\cdot}$). The formed peroxyacetyl radical finally reacts with nitrogen dioxide (NO_2) to generate peroxyacetylnitrate (PAN). The steps involved in the last sequence of reactions from the aldehycle are shown in chart (4.2).

$$\underset{\text{Aldehyde}}{RCHO} \xrightarrow{H\dot{O}} \underset{\text{Acyl radical}}{RC\dot{O}} \xrightarrow{O_2} \underset{\text{Peroxyacyl radical}}{RCO\dot{O}_2} \xrightarrow{NO_2} \underset{\text{Peroxyacetylnitrate (PAN)}}{R{-}COO_2NO_2}$$

Chart 4.2. The conversion of aldehyde to PAN

It is the peroxyacetylnitrate (PAN), a constituent of smog, which produces irritation to eyes. The two components of smog, viz, ozone and PAN affect the respiratory tract of human beings. It also causes nose and throat irritation and can lead to several problems of eyes, lungs and heart. The photochemical smog also affects the growth of plants and cause damage to the vegetation.

The formation of photochemical smog can be suppressed by preventing the release of oxides of nitrogen and hydrocarbons in the atmosphere by motor vehicles by using catalytic converters. The same objective can be achieved by using a better quality of oil, so that there is no emission of the oxides of nitrogen and hydrocarbons. This is possible by the use of so called **Reformed gasoline**, in which about 10% of the aromatic hydrocarbons are replaced by fuel oxygenates, such as methyl tertiarybutyl either. The additives make the fuel burn more efficienty. With this, the emission of volatile organic hydrocarbons is supressed. Also, the gasoline containing oxygenates burns at a lower temperature resulting in the reduction of the emission of the oxides of nitrogen. Also, with the advancement of technological developments, it is now possible to fabricate automobile engines so that there is no urburnt hydrocarbons left. This technology has been used in most of the advanced countries.

4.6 AIR POLLUTION

Air is invariably impure and is contaminated with gases like oxides of carbon (particularly carbon monoxide), nitrogen and sulphur (which are poisonous in nature) and finally divided into solid and liquid particles and smog. The presence of above contaminants causes air pollution. The air pollution, in fact, is an unwanted change in the quality of air caused by emission of gases due to burning of fossil fuels. Volcanic eruptions and discharge of particulate matter into the atmosphere by industrial activities and soil erosion.

Air is invariably impure and is always contaminated with gases like CO, NO_2, SO_2 etc (which are poisonous in nature) and finely divided solid and liquid particles and smog. Air becomes polluted due to the presence of the above contaminants. The presence of these contaminants in the air is called **air pollution** and the materials which pollute the air are called air pollutants. In fact, pollution of air (or atmosphere) is an unwanted change in the quality of air caused by emission of gases due to the burning of fossil fuels, volcanic eruptions and discharge of particulate matter into the atmosphere by industrial activities and soil erosion.

On a global scale, large number of deaths have been reported. These deaths are attributed to pollutants like oxides of sulphur and nitrogen and particulate matter which combine in presence of moisture to form smog, which in turn is responsible for respiratory diseases. In order to avoid the mishaps. Various governments have presented pollution standards for petrol, CNG/LPG and diesel vehicles. The quality of air has improved due to use of non-leaded petrol with low sulphur content.

Air pollution is also responsible for acid rain, global warming and formation of smog.

4.7 MAJOR AIR POLLUTANTS

The major air pollutants are those which are produced in significant amounts and are responsible for a variety of health and environmental problems. Table 4.1 below describes various air pollutants.

Table 4.1 Different Types of Air Pollutants

Type of Pollutant	*Examples*
Gaseous Pollutants	Oxides of Carbon (CO, CO_2) Oxides of Nitrogen (NO, NO_2) Oxides of Sulphur (SO_2, SO_3)
Particulate Pollutants	Smoke, dust particles
Smog and Photochemical Smog	A colloidal dispersion of gaseases pollutants like SO_2, NO_2 etc and water droplets with particulate matter
Metallic Pollutants	Metallic particles like Pb, Hg, As, Cd, Be, Ni, V, Fe, Sn, and Al etc released in the air by various industrial processes
Radiation	Released in the atmosphere by various nuclear devices etc.

Besides the above pollutants, pollution of the air is also caused by various chemicals discharged from industrial processes, petroleum, minerals and chlorofluorocarbons, etc.

4.7.1 Oxides of Carbon

Carbon forms two oxides, viz., carbon monoxide and carbon dioxide. The former (carbon monoxide) is one of the most harmful pollutants. The later (i.e. carbon dioxide) is harmless but in larger amounts leads to a number of problems.

Carbon Monoxide (CO)

Carbon monoxide is a colourless, odourless, tasteless and non-irritating gas. It is sparingly soluble in water and is highly toxic (poisonous) in nature. It reduces the oxygen carrying capacity of blood and can be fatal in concentrations exceeding 100 ppm. It is believed to be a silent killer. Major sources of carbon monoxide in the atmosphere are as follows:

(*i*) It is produced by the incomplete combustion of all carbon containing fuels. The incomplete combustion means combustion taking place in presence of insufficient quantity of oxygen, such combustion takes place in automobile engines (internal combustion engines).

(*ii*) Industrial furnaces (e.g., blast furnace) which are used at high temperature (e.g., in the manufacture of iron).

(*iii*) Cigarette smoke also contains carbon monoxide. Though a single smoke emits very small amount of carbon monoxide, but considering the number of people smoking a number of cigrettes everyday discharge considerable amount of carbon monoxide into the atmosphere.

(*iv*) Industrial gases also contain carbon monoxide. For example, coal gas, obtained by destructive distillation of coal contains 5–10% carbon monoxide. Also, water gas, obtained by passing steam over red hot coke ($H_2O + C \rightarrow CO + H_2$) contains 30 – 40% carbon monoxide.

(*v*) Besides the above sources, carbon monoxide is also produced from some natural sources like:

- Volcanic eruptions
- Forest fire
- Methane, an important constituent of natural gas and produced in marshy places on oxidation under controlled conditions give carbon monoxide.
- Decomposition or degradation of chlorophyll, the green pigment of green plants.
- Marine algae and marine siphmophore produce carbon monoxide.

Carbon Monoxide Poisoning

We know that during breathing, the haemoglobin (Hb) present in the blood combines with oxygen to form oxy-haemoglobin (HbO_2)

$$Hb + O_2 \rightleftharpoons HbO_2$$

As a result of blood passing through the arteries to the tissues, the pressure of oxygen decreases and the oxygen present in the complex HbO_2 is set free, where it is utilised by the cells.

In case, the air (we breathe) contains carbon monoxide, the oxygen present in HbO_2 is replaced by CO, since haemoglobin (Hb) has more affinity for CO than for O_2; the result is the formation of carboxy–haemoglobin (HbCO).

$$HbO_2 + CO \longrightarrow HbCO + O_2$$

Due to the formation of HbCO, the quantity of oxygen available to the body cells get reduced. This implies that carbon monoxide reduces the oxygen carrying capacity of the blood and this leads to anoxia (oxygen starvation). The deficiency of oxygen produces headache, dizziness, choking, cardiac and pulmonary complication leading to paralysis and death.

In case, carbon monoxide poisoning is detected well in time (i.e., before the person dies), it can be treated by exposing the victim to oxygen under a high pressure of 2 2.5 atm. By this process, the CO present in HbCO is replaced by O_2 and the victim may be revived.

$$HbCO + O_2 \rightleftharpoons HbO_2 + CO$$

$$HbO_2 \longrightarrow Hb + O_2$$

Determination of the concentration of carbon monoxide in air

The concentration or the amount of carbon monoxide is generally determined by non dispersive infrared analysis. It depends on the fact that carbon monoxide, which possesses a dipole moment, absorbs infrared radiations at certain specific wavelengths. The amount of infrared radiations absorbed is directly proportional to its concentration. This method is used if the level of carbon monoxide is up to a level of 150 ppm. In case the concentration of the gas is about 10 ppm, it can be estimated by gas chromatography, which makes use of flame ionisation detector. However, if the concentration of carbon monoxide is less than 10 ppm, it is estimated by reducing with hydrogen gas over a nickel catalyst at 360°C and the formed methane is estimated by flame ionisation detector.

$$CO + 3H_2 \xrightarrow[360^\circ C]{Ni} CH_4 + H_2O$$

Carbon Dioxide (CO_2)

Carbon dioxide is a colourless, odourless gas and is about 1.5 times heavier than air. It is non-poisonous and does not support life. However, animals and human beings if kept in a atmosphere of carbon dioxide die due to lack of oxygen leading to suffocation. Small amounts of CO_2 is considered relatively innocuous and so is not considered to be a pollutant. However, in excess amounts in the atmosphere, it has adverse effects on the climate (see greenhouse effect and global warming). So in large amount, carbon dioxide is considered to be a pollutant. Major sources at carbon dioxide in the atmosphere are as follows:

(*i*) Burning of fossil fuels like coal, natural gas and petroleum etc produce carbon dioxide. It is estimated that about 6000 million tons of CO_2 enter the atmosphere every year by such activities.

(*ii*) The cultivation of soil also releases large amounts of CO_2, which is produced by bacteria. This activity releases more than 2000 million tonnes of CO_2 every year into the atmosphere.

(*iii*) Industrial process like manufacture of lime, alcohol and other fermentation units contribute more than 10,000 million tonnes of CO_2 into the atmosphere

$$\underset{\text{Lime stone}}{CaCO_3} \xrightarrow{1250^\circ} \underset{\text{Lime}}{CaO} + CO_2$$

$$\underset{\text{Glucose}}{C_6H_{12}O_6} \xrightarrow{\text{Zymase}} \underset{\text{Ethyl alcohol}}{2C_2H_5OH} + 2CO_2$$

(*iv*) Decay of dead organisms, respiration by living organisms, volcanic eruptions and forest fires and some other activities acid also release CO_2 in the atmosphere.

The major consumers of CO_2 are the green plants, which absorb CO_2 from the atmosphere in presence of sunlight and chlorophyll to produce their food in the form of carbohydrates. This is known as **photosynthesis.**

$$nCO_2 + nH_2O \xrightarrow[\text{Chlorophyll}]{\text{Sunlight}} \underset{\text{Carbohydrate}}{(CH_2O)_n} + nO_2$$

Oceans also dissolve CO_2 from the atmosphere to form carbonate rocks. In this way both the plants and oceans keep a balance of CO_2 in the atmosphere.

Effects of CO_2 in the atmosphere

The main effects of CO_2 in the atmosphere are its contribution in the green house effect, global warming and in maintaining the pH of rain water at about 5.6. All these form the subject matter of a subsequent discussion (see Section 4.9).

4.7.2 Oxides of Nitrogen

Nitrogen combines with oxygen to form a number of oxides. These are nitric oxide (NO), nitrogen dioxide (NO_2), nitrogen trioxide (N_2O_3), nitrogen tetraoxide (N_2O_4) and dinitrogen pentoxide (N_2O_5). From the viewpoint of environmental pollution, the first three oxides, viz., nitric oxide, nitrogen dioxide and nitrous oxide are important and are discussed here. Nitrous oxide is also a green house gas (see greenhouse effect). It plays an important role in the destruction of ozone layer. Major sources of oxides of nitrogen in the atmosphere are as follow:

(*i*) The atmospheric nitrogen and oxygen combine during thunderstorm and lightning to form nitric oxide, which gets converted into nitrogen dioxide.

$$N_2 + O_2 \xrightarrow[\text{discharge}]{\text{electric}} 2NO$$

$$2NO + O_2 \longrightarrow 2NO_2$$

(*ii*) The exhaust gases released from automobiles contain NO and NO_2 besides other gases.

(*iii*) Nitrogen dioxide is obtained as a by-product during the manufacture of explosives like TNT.

(*iv*) Bacterial decay of organic matter by soil micro organisms release oxides of nitrogen into the atmosphere.

(*v*) Nitrous oxide is produced by the action of aerobic and anaerobic bactria in the soil on nitrogen based fertilizers.

Functions of oxides of nitrogen in the atmosphere

As already stated, nitrogen and oxygen combine in the atmosphere during thunderstorm and lightening to form nitric oxide. The nitric oxide formed nitrogen dioxide by aerial oxidation.

$$N_2 + O_2 \longrightarrow 2NO$$

$$2NO + O_2 \longrightarrow 2NO_2$$

Nitric oxide can also be oxidised to nitrogen dioxide with ozone.

$$NO + O_3 \longrightarrow NO_2 + O_2$$

Nitrous oxide (which is produced by the action of bacteria in the soil on nitrogen based fertilizers) is unreactive in the lower atmosphere. However, in the stratosphere, it (N_2O) gives nitric oxide on photolysis

$$N_2O + h\nu \longrightarrow NO + N$$

Nitrous oxide may also give nitrogen and oxygen on photolysis.

$$N_2O + h\nu \longrightarrow N_2 + O$$

The free oxygen atom may react with nitrous oxide producing nitric oxide, which is subsequently oxidsed to nitrogen dioxide.

$$N_2O + O \longrightarrow 2NO$$

$$NO + O \longrightarrow NO_2$$

In the atmosphere, NO_2 reacts with water vapor producing HNO_3, which is washed down as acid rain (see acid rain, section 4.9.1).

$$3NO_2 + H_2O \longrightarrow 2HNO_3 + NO$$

Alternatively, nitric acid may combine with ammonia (present in the atmosphere) giving ammonium nitrate

$$NH_3 + HNO_3 \longrightarrow NH_4NO_3$$

The acid rain on coming in contact with marble ($CaCO_3$) of monuments or lime stone gives CO_2 gas and calcium nitrate.

$$CaCO_3 + 2HNO_3 \longrightarrow Ca(NO_3)_2 + CO_2 + H_2O$$

In the atmosphere, nitric oxide destroys the ozone (which absorbs harmful uv radiation coming from the sun).

$$NO + O_3 \longrightarrow NO_2 + O_2$$

$$NO_2 + O_3 \longrightarrow NO + 2O_2$$

Thus both nitric oxide and nitrogen dioxide destroy ozone layer (see chapter 4).

Nitrogen dioxide is harmful for plants and causes extensive leaf-drops. It has been found that inhalation of NO_2 (15 – 30 ppm concentration) for about 2 hours, causes damage to lungs, liver and kindly.

Oxides of nitrogen on reaction with hydrocarbons and sunlight lead to the formation of photochemical smog.

Determination of the concentration of the oxides of nitrogen in the atmosphere

In the atmosphere, both nitric oxide and nitrous oxide are finally converted into nitrogen dioxide. So the determination of concentration of oxides of nitrogen is done by determining the concentration of nitrogen dioxide. The procedure involves, bubbling the contaminated air (containing NO_2) into a solution of NaOH for a period of 24 hr. The NO_2 in the contaminated air is converted into $NaNO_2$ and $NaNO_3$.

$$2NO_2 + 2NaOH \longrightarrow NaNO_2 + NaNO_3 + H_2O$$

The above solution is added to a mixture of hydrogen peroxide, sulphanilic acid, hydrochloric acid and N-1(naphthyl) ethylenediamine dihydrochloride. The sodium nitrite formed in the reaction diazotises sulphanilic acid. This procedure is carried out at low temperature (0 – 5°). The formed diazonium salt couples with N-(1-naphthyl) ethylenediamine dihydrochloride to give the coupled compound. The reactions involved are as follows:

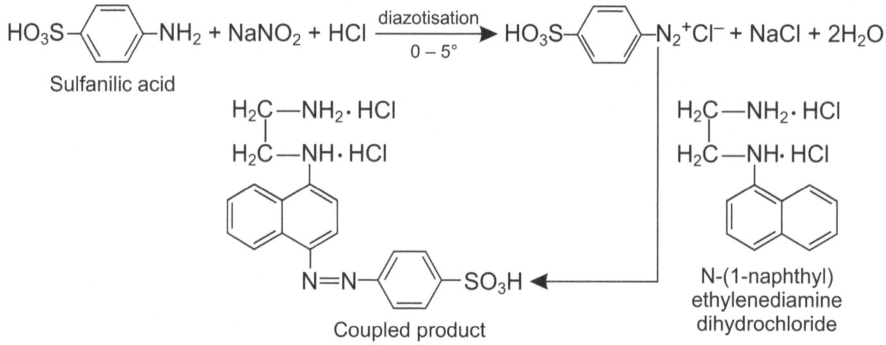

In an alternative procedure, the oxides of nitrogen in the atmosphere are made to react with ozone to yield electronically excited species ($N\dot{O}_2$).

$$NO + O_3 \longrightarrow N\dot{O}_2 + O_2$$

$$NO_2 + O_3 \longrightarrow NO + 2O_2$$

$N\dot{O}_2$ being unstable quickly returns to the ground state, forming NO_2 and emitting light is the range of 600 – 3000 nm.

$$N\dot{O}_2 \longrightarrow NO_2 + h\nu$$

The intensity of light is directly proportional to the concentration of $N\dot{O}_2$ and hence NO molecules. The intensity of light, measured with the help photo-multipliers give a measure of the concentration of NO molecules.

4.7.3 Oxides of Sulphur

Sulphur forms two oxides, viz., sulphur dioxide and sulphur trioxide. The former oxide, viz., sulphur dioxide is important from the point it view of atmospheric pollution.

Sulphur Dioxide (SO_2)

It is a colourless gas with suffocating smell. It irritates the respiratory system and damages lungs and so is a harmful gaseous pollulant. The main sources of SO_2 is the atmosphere are as follows:

(*i*) Volcanic eruptions.

(*ii*) Both oil and coal contain sulphur as impurity and on burning release SO_2 into the atmosphere.

(*iii*) Roasting of sulphlide ones, like pyrites (FeS), copper pyrites (CuS), copper glance (Cu_2S), zinc blende (ZnS), galena (PbS) etc in air during the metallurgy of respective metals release considerable amounts of SO_2 in the atmosphere.

(*iv*) Power plants, oil or coal based, oil refineries and H_2SO_4 manufacturing plants, release SO_2 into the atmosphere.

(*v*) Hydrogen sulphide discharged into the air by natural processes is converted into SO_2 by oxidation in the atmosphere.

Effects of oxides of sulphur in the atmosphere

In the atmosphere, SO_2 reacts with oxygen to form SO_3 by photolytic and catalytic oxidation processes

$$SO_2 + \frac{1}{2}O_2 \xrightarrow{h\nu} SO_3$$

$$SO_2 + \frac{1}{2}O_2 \xrightarrow{\text{Particulars}} SO_3$$

Sulphur dioxide also reacts photo-chemically with ozone producing SO_3.

$$SO_2 + O_3 \xrightarrow{h\nu} SO_3 + O_2$$

The SO_3 formed in the atmosphere reacts with water forming H_2SO_4

$$SO_3 + H_2O \longrightarrow H_2SO_4$$

The sulphuric acid formed comes down to the earth as acid rain (see section 4.9.1). The acid rain damages the crops and may react with ammanium salt and sodium chloride in the soil to produce sulphates.

$$H_2SO_4 + 2NH_4Cl \longrightarrow (NH_4)_2SO_4 + 2HCl$$

$$H_2SO_4 + 2NaCl \longrightarrow Na_2SO_4 + 2HCl$$

Sulphuric acid also damages marble (as in the case of HNO_3)

$$H_2SO_4 + CaCO_3 \longrightarrow CaSO_4 + H_2O + CO_2$$

Prolonged exposure of plants to even small concentration of SO_2 results in decolorisation of leaves. This is because the production of chlorophyll is hampered. The colour of fabrics, leather, paper are paints fade in presence of SO_2.

Determination of the concentration of SO_2 in the air

The concentration of sulphur dioxide in the air can be determined by bubbling test sample of air through a dilute aqueous solution of sodium tetrachloromercurate (II), $Na_2[HgCl_4]$ (obtained by the action of $HgCl_2$ on NaCl solution) [$HgCl_2$ + 2 NaCl $\rightarrow$ $Na_2[HgCl_4]$] containing sulphamic acid (which destroys any nitrogen oxides present in the air). Sulphur dioxide in the sample of air reacts quantitatively with sodium tetrachloromercurate (II) and is converted into a stable complex, sodium dichlorosulplhitomercurate (II), $Na_2[HgCl_2(SO_3)]$.

$$Na_2[HgCl_4] + SO_2 + H_2O \longrightarrow Na_2[HgCl_2(SO_3)] + 2\,HCl$$

The above complex is treated with a mixture of *p*-rosaniline dye, formaldehyde and phosphoric acid. The *p*-rosaniline dye is converted to *p*-rosaniline methyl sulphonic acid.

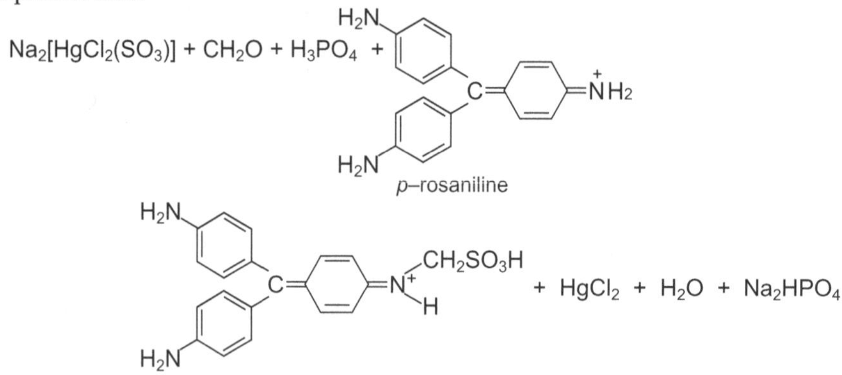

p–rosaniline methylsulphonic acid
(red–violet colour)

In the above procedure, phosphoric acid liberates SO_2 from the complex and maintains the pH of the reaction at 1.0; it also ties-up any heavy metal ions that may be present in the environment along with SO_2.

The *p*-rosanilinemethylsulphonic acid has a red-violet colour and at pH 1.0, it absorbs at 575 nm. Its concentration can be determined calorimetrically using Lambert-Beer's law. This concentration is proportional to the concentration of SO_2 in the atmosphere.

4.7.4 Particulates

The finely divided suspended solid particles in the atmosphere are called *particulates*. The diameter of the particulates lies in the range 0.001 – 100 μm and cause severe health problems. Some sources of particulates are as follows:

(*i*) Smoke emitted into air by the process of combustion of fuels like coal and oil in homes and factories. Power generation plants which use coke are the greatest source of smoke. The smoke basically contains particles of carbon, which remain suspended in air.

(*ii*) Cement dust emitted from cement factories

(*iii*) Insecticide dust

(*iv*) Coal dust

(*v*) Fly ash generated from power units

Adverse Effects of Particulates

Particulates produce a number of adverse effects which are as follows:

(*i*) Particulates play a major role in the formation of photochemical smog.

(*ii*) Particulates aggravate the deleterious effects of other air pollutants. The particulates sometimes travel across the national boundaries and so the harmful effects are felt many thousands of kilometers from their place of origin.

(*iii*) Women exposed to fine particulate matter (having diameter less than 10μm) have been found to deliver children having small heads and bodies. Such children are slow to learn and have an increased risk of developing cancer. It is believed that particulates which are coated with polynuclear hydrocarbons, damage the DNA of the growing fetus; the damage caused cannot be reversed.

(*iv*) Visibility is considerable reduced.

(*v*) Mercury if present as particulate, produces heaviness, headache, nervousness, fatigue and a number of other problems. Prolonged exposure causes the breakdown of CNS.

(*vi*) *Automobiles*: The exhaust gases from automobiles discharge account to the extent of about 1 per cent of the total mass of particulates in the atmosphere. These particulates comprise of soot, lead compounds, crank case oil and hydrocarbons (besides other gaseous pollutants like CO, CO_2, oxides of nitrogen and sulphur etc).

(*vii*) *Combustion processes*: The main components of various combustion processes are SO_2 and NO_2. These are converted into sulphates and nitrates via the formation of sulphuric acid and nitric acid. Incomplete combustion of fossil fuels produces soot particles. A number of combustion process produce fly ash which is a mixture of oxides of silicon, aluminium, calcium and phosphorus.

(*viii*) *Metallurgical processes*: These processes emit fine metallic dust and oxides of metals which provide catalytic surface for a number of unwanted reactions in the atmosphere.

(*ix*) The presence of lead as particulate is extremely harmful. It is a cumulative poison and keeps on accumulating in the tissues of the human body and in the leaf tissues of plants. Accumulation of lead in human tissues causes malfunctioning of red blood cells leading to anaemia. Lead is known to damage organs like liver, kidneys and intestines and affects the CNS.

(*x*) Cadmium pollution may arise from the metallurgical processes of zinc, and copper. It causes lung irritation, vomiting and hypertension.

Following table lists some more metallic pollutants and their effects on human beings.

Table 4.2 Sources of Some Metallic Pollutants and their Effects.

S.No. Metallic pollutant	***Major sources***	***Harmful effects***
i. Zinc (Zn)	Zinc refineries, galvanising processes, brass manufacture, metal plating, and plumbing.	Zinc fumes have corrosive effect on skin and can cause irritation and damage mucous membranes.
ii. Chromium (Cr)	Metallurgical and chemical industries, processes using chromate compounds, cement and asbestos units.	Toxic to body tissues, can cause irritation, dermatitis, ulceration of skin, perforation of nasal septum. Carcinogenic action suspected.
iii. Arsenic (As)	Arsenic containing fungicides, pesticides and herbicides, metal smelters, byproduct of mining activities, chemical wastes.	Inhalation, ingestion or absorption through skin can cause mild bronchitis, nasal irritation or dermatitis. Carcinogenic activity also is suspected. Attack-SH groups of enzymes,coagulate proteins.
iv. Beryllium (Be)	Coal, nuclear power and space industries, production of fluorescent lamps, motor fuels and other industrial use.	Damage to skin and mucous membranes, pulmonary damage, perhaps carcinogenic.
v. Boron (B)	Boron producing units, production and use of petroleum fuel and additives, burning coal and industrial wastes, detergent formulations.	Ingestion, inhalation as dust causes irritation and inflammation. Boron hydrides can damages CNS and may result in death.
vi. Manganese (Mn)	Ferromanganese production, organo-manganese fuel additive, welding rods, incineration of manganese containing substances.	Absorption, ingestion, inhalation, or skin contact may cause manganic pneumonia.
vii. Nickel (Ni)	Metallurgical industries using nickel, combustion of fuels containing nickel additives, burning of coal and oil, electroplating units using nickel salts, incineration of nickel containing substances, vanaspati manufacture.	Respiratory disorder, dermatitis, cancer of lungs and sinus.

(Table 4.2 Contd...)

S.No. Metallic pollutant	Major sources	Harmful effects
viii. Vanadium (V)	Vanadium refining, production of vanadium containing alloy's power plants, burning of oil rich in vanadium.	Gastrointestinal and respiratory disorders inhibition of synthesis of cholesterol, heart disease and cancer in case of chronic exposures.
ix. Selenium (Se)	Burning of fuels and residual oils, fumes and gases from refinery wastes, incineration of paper and other wastes natural sources.	Irritation of gastrointestinal and respiratory tracts, irritation of eyes, nose and throat, damage to lungs, liver and kidneys.

4.7.5 Smog and Photochemical Smog

Smog is a mixture of smoke and fog. It is two types: classical smog and photochemical smog. For details see sections 4.5.1 and 4.5.2

4.7.6 Metallic Pollutants

The wastes from metal industries usually contain metals like Cu, Pb, Cd, Cr, Zn, As, etc. These are highly toxic to human beings and aquatic life. Also see Table 2.20 and 4.2.

4.7.7 Radiation

Radiation results from radioactive wastes, radioactive fall outs, radioisotopes, natural radioactive nuclides and nuclear accidents. For details see page 84-85.

4.7.8 Chemicals

The wastes from chemical industries like alkali manufacturing industries, fertilizer industries, pesticides and a large number of industries that manufacture organic as well as inorganic chemicals are responsible for polluting the land with various chemicals. The industrial wastes are mostly acidic and so it is necessary to be neutralised before discharge. The wastes from rayon manufacturing industries and DDT manufacturing industries are highly acidic and corrosive.

The wastes from the production of chemicals used to make insecticides, and hersecides, pesticides arise from the production of 2.4 dischlorophenoxy-acetic acid (2.4.D) are classified as dangerous wastes. All these wastes if dumped into the land find their ways into ground water and water bodies by rains etc and pollute these bodies.

4.7.9 Petroleum

Petroleum is naturally occurring oil, mostly found in underground reservoirs. It is believed to have been formed by the decay of the remains of animals and plants. It is also found below the oceans and is believed to be formed by decay of marine plants and animals. In its crude form, petroleum is a mixture of hundreds of chemicals, chiefly hydrocarbons. It is separated into various products such as cooking gas, naphtha, petrol, diesel, kerosene and wax by fractional distillation in refineries. The pollution of air or atmosphere occurs at different stages. Some of these are:

- During drilling, some of low boiling fractions escape in the atmosphere and spill on the land. In case, drilling from below the surface of oceans, the petroleum oil may be spilled on the surface of water and may be swept by air currents to distant places.
- During transportation of crude petroleum oil, there is spilling or leakage of petroleum. This particularly occurs in oceans during transportation by huge tankers.
- During fractional distillation in the refineries, a number of fractions, e.g., gas, hydrocarbon, etc., find their way into the atmosphere due to leaks or poor recovery.

4.7.10 Chlorofluorocarbons

The chlorofluorocarbons are compounds containing carbon, chlorine and fluorine. The CFC's are non-toxic, non-inflamable and are used as refrigerants (for refrigerators and air-conditions). Though these are unreactive, the CFCs are a major cause of ozone depletion into the stratosphere. (see section 4.9.4).

4.8 SOURCES OF AIR POLLUTION

The air pollution is caused by carbon monoxide, carbon dioxide, oxides of nitrogen, oxides of sulphur, hydrogen sulphide, chlorine, ozone, hydrocarbons and particulates. All these have been described in various sections. The present Section describes some of the other major sources of air pollution. These include pollution from

- Industries
- Petroleum
- Mining industries
- Power plants
- Automobiles

Air pollutants are classified as either directly released or formed by subsequent chemical reaction. Air direct released air pollutant is one that is emitted directly from a given source, such as the carbon monoxide or sulfur dioxide, all of which are byproducts of combustion; whereas, a subsequent air pollutant is formed in the atmosphere through chemical reactions involving direct release pollutants. The formation of ozone in photochemical smog is the most important example of a subsequent air pollutant.

Pollutants that are directly released into the atmosphere as a result of natural or other activities are called primary pollutants. Examples include oxides of nitrogen and sulphur and particulate matter on the other hand the primary pollutants generate other pollutants called secondary pollutants. Examples includes formation of sulphuric acid (from SO_2, O_2, H_2O) and nitric acid (from oxides of nitrogen, H_2O). Following are given some of the sources of pollution.

- Forest fires generally started by lightning
- Soil erosion
- Volcanic eruptions
- Discharge of organic compound from leaves and trees
- Putrification of organic matter
- Natural radioactivity
- Effulants from industries like polymer and plastics, asbestos, extraction and refining of petroleum, Mining industries (like coal, asbestos, metals, mica, silica, etc)
- Burning of fossil fuels as in automobiles, buses, aircrafts using gasoline or diesel
- Generation of Power by burning of coal
- Extraction of metals from ores
- Manufacture of cement
- Testing of nuclear devices
- Biomedical waste from hospitals.

4.8.1 Pollution from Industry

It has already been stated that industrial units dealing with the manufacture of explosives (various nitro compounds like TNT) using concentrated nitric acid and concentrated sulphuric acid release thousands of tones of oxides of nitrogen and oxides of sulphur into the atmosphere. Industries dealing with fermentation technology (e.g., in the manufacture of antibiotics and other chemicals like ethyl alcohol) release thousands of tonnes of carbon dioxide into the atmosphere. Caustic soda manufacturing units release thousands of tonnes of chlorine in the atmosphere. Various metallurgical processes are responsible for introducing carbon monoxide, carbon dioxide, hydrogen sulphide and metal particulates into the atmosphere. Industrial pollution from some of the industries form the subject matter of chapter 5. The following discussions include some very special industries dealing with highly specialised products.

Polymer and Plastics

The polymers and plastics are nonbiodegerable. This has caused series waste disposal problems. For details see chapter 3 (Pages 94-98).

Asbestos

The name asbestos is given to a group of silicate minerals that can be separated into fibres and used to make materials that will not burn. The name asbestos comes from a Greek word meaning *inextinguishable* or *unquenchable*. It is the only naturally occurring fibrous mineral. It finds a variety of uses and is valued as an industrial material. Due to its fibrous nature, asbestos is an environmental hazard. Asbestos fibres are readily absorbed on the surface of lung tissues; this results in a disease called *asbestosis* and is also responsible for development of lung cancer. In fact, industrial workers who have worked in asbestos industries for about 20 years or so are susceptible to be affected by asbestosis as well as cancer.

Asbestos has been in use for more than 3000 years in commercial, public and other facilities. It is used because of its incombustibility and effectiveness as a reinforcing agent when combined with other materials like cement or plastic. Its resistance to chemical and biological attack is of great value. Finally, asbestos is cheap and easily available. Some important applications of asbestos are as follows:

- The fibres of asbestos can be spun into yarn, which can be reinforced with other fibres, e.g., nylon. Such yarn is used for making clothing for fire fighters and special gloves.
- Thermal insulators
- Electrical insulators
- For making brake lining of automobiles
- Building materials

Sources of Asbestos in the Environment

Asbestos present in some rocks and minerals which is not exploited may find its may into the atmosphere by natural phenomena as weathering, landslides etc. Workers engaged in mining, milling and fabrication processes are also exposed to asbestos. When the building materials containing asbestos deteriorate, the asbestos particles are released into the atmosphere. The airborne asbestos is finally ingested through inhalation. Likewise, asbestos is also released into the atmosphere by used brake lining of vehicles.

Monitoring of Asbestos in the Atmosphere

The concentration of asbestos in the atmosphere is determined as given below:

(*i*) The airborne asbestos particles are colloidal in nature and bear electrostatic charge. The first step involves the discharge or removal of the charge. For this, a sample of air (about 100 m^3) is impinged on the plate of a electrostatic precipitator.

(*ii*) The particles of asbestos settled on the plate, are washed into about 100 mL of a suitable non-polar solvent which dissolves unwanted organic particulates.

(*iii*) The undissolved asbestos particles are filtered.

(*iv*) The number of asbestos particles is finally counted under an electron microscope. The concentration is reported as number of asbestos particles per mL of air.

Harmful Effects of Asbestos

Exposure to asbestos fibres has the following effects:

(*i*) Incorporation of asbestos fibres in the lung tissues causes a lung disease known as **asbestosis**. Once inhaled, the asbestos particles get embedded in the lung tissues for long periods (20–30 years) and the victim's outer lining of the lungs gets thickened; thereby, decreasing the oxygen-carbon dioxide exchange capacity and the elasticity of the lungs. The victim becomes short of breath and eventually dies of heart failure. It is found that workers engaged in asbestos industries (e.g. mining, cleaning and weaving) are more prone to suffer from asbestosis. This disease is incurable.

(*ii*) **Cancer** Workers in industries dealing with the use of asbestos such as, in building, construction, mining, manufacture of brakes and repair of clutches of automobiles face 50% chances of dying from cancer. Since cancer symptoms do not appear (or arise) until 20 – 30 years after being exposed to asbestos, various agencies could not recognise the dangers posed by the use of asbestos. When inhaled, the fine particles of asbestos dust are lodged in the lungs and damage the cells therein. Prolonged heavy exposure increases a person's risk of lung cancer and *mesothelioma*, a rare cancer of thin membrane that line chest and abdomen. Some developed countries have banned the use of asbestos in any form. However, some developing and under developed nations have still not banned the use of asbestos completely.

(*iii*) **Iron metabolism** Certain varieties of asbestos contain appreciable amount of iron (25 – 26%). Though iron is essential for life, its excessive built up is harmful. In the body, iron forms chelate with citrate ion; these chelates enter the cells and ferrous iron reacts with oxygen producing molecular oxygen radicals ($O_2^{\cdot}$). These radicals create havoc; these oxidise DNA and proteins and cause damage to proteins. This leads to carcinogenesis.

Checking Asbestos Pollution

(*i*) Asbestos is most harmful in the form of dust. It is not possible to prevent atmospheric pollution, but one can take adequate precautions so that one does not inhale asbestos dust as far as possible. For example, workers who deal with asbestos must wear masks in order to prevent inhalation.

(*ii*) Asbestos sheets, if used must be coated with dust-proof membranes, such as paints or plastics or keeping the sheets wet while handling; this reduces the hazard. One should not be exposed to asbestos dust for long.

(*iii*) Asbestos dust arising out of cleaning brake lining should be cleaned by vacuum extraction procedure or with a damp rag.

(*iv*) Children are more susceptible to asbestos pollution than adults. This is especially important because in old buildings of schools, the roofs and tiles are made from asbestos. It is best to find such buildings and the places in which asbestos is present and these must be coated with polyvinyl chloride spray.

(*v*) As far as possible, the use of asbestos should be avoided. In its place, clothing made from nylon and wool blends (these provide heat and fire protection) should be used. For insulation, ceramic fibres can be used as a substitute for asbestos. Moulding of phenol-formaldehyde can be used in place of asbestos in electrical insulation materials. Also, glass fibre reinforced cement is a good substitute for asbestos based building materials.

Tobacco

The use of tobacco is responsible for 3 out of every 10 cancer deaths. It is known to be a major factor for cancer of the lung, larygeal, oral, esophageal, urinary bladder, and kidney and prostrate. Tobacco smoke has been found to contain over 2500 chemical substances. It contains carbon monoxide, carbon dioxide, ammonia, nitrosamines, nitrogen oxides, hydrogen cyanide, sulphur compounds, nitrites, ketones, alcohols and acrolein. The 'tars' contain carcinogenic hydrocarbons, which include nitrosamines, benzo(a)pyrene, anthracene, acridines, quinoline, benzene, naphthol, naphthalene, cresols and insecticides (DDT) as well as radioactive compounds like potassium-40 and radium-226. Some of these compounds are known to be risk factors for human cancer.

Cigarette smoking has been shown to be a major risk factor in men and women especially in middle age for cardiovascular diseases. Chronic lung diseases related to smoking are second to heart diseases as a cause of disability. Smoking adversely affects fertility in females. Children born to smoking mothers are more prone to various abnormalities.

The cigarette manufacturing industries all over the world make an unimaginable number of cigarettes all of which are smoked. One cannot even imagine the amount of total smoke released into the atmosphere with all the harmful substances (chemicals). In fact, smoking is a major factor contributing to atmospheric pollution.

The most important question that arises is how to control pollution caused by smoking. This has to be a voluntary effort of all the people who smoke to quit smoking. Can one imagine a day when cigarette smoking is considerably reduced so that the cigarette-manufacturing companies find it non-profitable to manufacture cigarettes? The answer is probably 'NO'. The reason is that the society we live in, it is not possible to have a law to ban smoking or the manufacturing of cigarettes.

Petroleum

Petroleum is found in underground reservoirs and below the oceans. It is a major cause of pollution particularly marine pollution. for details see section 4.7.9.

Mining Industries

Mining is the process that is used to take minerals out of the earth. Coal, asbestos, copper, mica, precious stones and many other materials are mined. Following mining industries contribute pollutants to the atmosphere.

(*i*) **Coal mining:** This discharges large amount of particulate matter in the form of coal dust into the atmosphere. Workers in coal mining are prone to be incapacitated by black lung disease.

(*ii*) **Mica mining:** Mica is a group of silicate minerals having layered structure. It contains mixed silicates of elements sodium, potassium, calcium, aluminium, magnesium, iron and lithium. It is mined from pegmatite rocks. India is the largest producer of mica in the world contributing more than 75 per cent of the world's consumption. In India, mica is found in Bihar, Andhra Pradesh and Rajasthan. Besides, it is also found in Rocky Mountains of USA and in Canada. Mica is widely used as an electrical insulator and in electrical capacitors. During mining and processing, mica is discharged into the atmosphere in the form of dust. This dust, on inhalation, causes a number of health hazards. The workers in the mining operations are the worst affected.

(*iii*) **Asbestos mining:** See page 125–128.

(*iv*) **Mining of ores:** An ore is naturally occurring mineral which is obtained by mining and is the raw material for the extraction or isolation of metals on a commercial scale. Most of the ores are compounds (in impure form) of metals with oxygen, sulphur, silicon or phosphorus. The pollution of air or atmosphere occurs at different stages. Some of these are as follows:

- During mining by drilling etc. fine particles of the ores are discharged into the atmosphere.
- During extraction or isolation of metals—the ores are subjected to various processes like concentration, roasting, smelting etc. In all these processes, gaseous pollutants like carbon monoxide, carbon dioxide, sulphur dioxide are introduced into the atmosphere. Some metals which are comparatively of low melting point vaporise and pollute the atmosphere.

Power Plants

Power plants are used for generating power or electricity. In most of the con-ventional power plants, the source of energy is coal. Coal fired power plants are a source of serious environmental problems. These inject the gaseous pollutants like carbon monoxide, carbon dioxide, sulphur dioxide (if coal contains sulphur impurities) into the atmosphere. Besides gaseous pollutants, such power plants inject particulates in the form of fine carbon particles and fly ash (the ash left after coal is heated). A number of methods have been used to reduce the pollution caused.

With the advancement of knowledge nuclear energy is used to produce electricity at a much cheaper rate. However, in this case, elaborate arrangements for generation and control have to be used. There are a number of problems associated with the nuclear energy production.

- There is deleterious effects on humans due to the release of ionising radiations.
- There is considerable problem in the disposal of the spent fuel.
- There may be accidents which may have disastrous effects on mankind.

Automobiles

Automobiles are mostly run on petrol or diesel oil. The exhaust gases from the automobiles inject a number of gaseous pollutants like carbon monoxide, carbon dioxide, oxides of nitrogen and sulphur dioxide. In fact, automobiles are major contributor of atmospheric pollutants. In case of automobiles, the pollutants are removed by passing the exhaust gases through a catalytic converter.

The aeroplanes use huge amounts of gasoline and are the major contributors of exhaust gases in the higher atmosphere.

4.9 EFFECTS OF AIR POLLUTION

Vanous air pollutants are responsible for the following:

- Acid Rain
- Green House Effect
- Global Warming
- Depletion of Ozone
- Formation of Smog

4.9.1 Acid Rain

The formation of nitric acid and sulphuric acid as secondary pollutants in the atmosphere results in a kind of pollution known as acid rain. It includes all types of rain and snow whose excessive acidity causes environmental problems like destruction of vegetation and marine life and the corrosion and etching of buildings that are exposed to the atmosphere.

Chemically speaking, all rain is acidic with or without air pollution. This is due to the natural presence of carbon dioxide in the atmosphere which dissolves in rain drops or rain water (even moisture present in the atmosphere does the same function) to form carbonic acid.

$$CO_2 + H_2O \longrightarrow H_2CO_3$$

Due to the above reaction, carbon dioxide can dissolve in water until the solution becomes saturated. This results in the rainwater attaining an acidic pH of 5.6. Due to this, the purest form of rain (in the absence of any of the air pollutants) reaches the earth as an acidic solution of pH 5.6. In view of the above, the acid rain is described as rain whose pH is lower than 5.6.

How is acid rain caused?

(*i*) Due to the presence of oxides to nitrogen in the atmosphere During thunderstorm and lightning, atmospheric nitrogen and oxygen combine to form nitric oxide, which gets converted to nitrogen dioxide, which reacts with water to form nitric acid and is washed down as acid rain (see also section 4.4.2 and 4.4.6).

$$N_2 + O_2 \xrightarrow[\text{discharge}]{\text{electric}} 2NO$$

$$2\,NO + O_2 \longrightarrow 2\,NO_2$$

$$3\,NO_2 + H_2O \longrightarrow 2\,HNO_3 + NO$$

The oxides of nitrogen are also injected into the atmosphere from other sources (section 4.7.2).

(*ii*) Due to the presence of sulphur dioxide in the atmosphere Sulphur dioxide is released into the atmosphere from a number of sources (see section 4.7.3). In the atmosphere, sulphur dioxide reacts with atmospheric oxygen to form sulphur trioxide by photolytic and catalytic oxidation processes. The formed sulphur trioxide reacts with water forming sulphuric acid which comes down on earth as acid rain.

$$SO_2 + \frac{1}{2}O_2 \xrightarrow{h\nu} SO_3$$

$$SO_3 + H_2O \longrightarrow H_2SO_4$$

(*iii*) The presence of hydrogen sulphide and chlorine in the atmosphere also contributes to the formation of acid rain.

The acid rain is not confined to the region where the harmful pollutants (like oxides of nitrogen and sulphur) are emitted. Since winds carry atmospheric contaminants across national borders, the acid rain may fall hundreds of miles away from the pollution sources. Acid rain, in fact, has become an international issue. It is the responsibility of all countries to make sure that the pollutants are not discharged into the atmosphere; these should be checked at the initial source only.

Harmful Effects of Acid Rain

Acid rain causes extensive damage to the environment.

(*i*) Vegetation: The acid rain makes the soil acidic. This adversely affects the plants and animals. The acid rain contains H^+, SO_4^{2-} and NO_3^- ions which when added to soil, leaches the nutrients from the soil. The acid rain also damages the leaves of plants and trees. This is responsible for retarding forests and other vegetation.

(*ii*) Fertility of Soil: The activity of symbiotic nitrogen fixing bacteria present in the nodules of leguminous family is inhibited. This is responsible for destroying or reducing the fertility of the soil.

(*iii*) Aquatic Life: Acid rain renders the river or even ocean waters acidic, thereby, adversely affecting marine animals. Changes in pH of fresh water affect the reproduction and survival of many species of fish.

***(iv)* Building and Monuments:** Acid rain causes extensive damage to buildings and monuments made from marble, limestone, slate and mortar. Limestone which is very common building material, is attacked readily.

$$CaCO_3 + H_2SO_4 \longrightarrow CaSO_4 + H_2O + CO_2$$

Most of the monuments are made from marble which is mostly $CaCO_3$ and so are attacked by acid rain.

The damage caused to the buildings and monuments by acid rain is irreversible. A most glaring example is slow degradation of Taj Mahal in Agra (UP, India). The pollutants in and near Agra arise from iron foundries, rubber manufacture, brick kilns, and oil refinery at Mathura.

***(v)* Ecological Balance:** Acid rain is responsible for wiping out many bacterial and blue green algae; thereby disrupting the whole ecological balance.

***(vi)* Human Health:** Acid rain has been found to be very dangerous to human health. Acidic conditions can play havoc within human nervous system, respiratory system and digestive system by making the person an easy prey to various neurological diseases.

It is of utmost importance to control the acid rain. Short-term control of acid deposition (soil) can be achieved by using lime.

4.9.2 Greenhouse Effect

It is the phenomenon in which the earth's atmosphere traps the heat form the sun and prevents it from escaping into outer space. Without this effect, the average surface temperature of the earth would be about 33°C lower than it is today. The greenhouse effect received its name because the earth's atmosphere acts much like the glass or plastic roof and walls of a greenhouse. Sunlight enters a greenhouse through the transparent glass of plastic and heat the plants but the heat emitted by the plants in the form of infrared radiation cannot pass through the glass or plastic roof and walls. As a result, the temperature inside increases.

In the atmosphere, there is a protective layer of O_3 gas at a height between 15 – 60 km (the thickest layer of ozone exists at a height of 23 km from the surface of the earth and a blanket of CO_2 gas exists in the lower part of the atmosphere (i.e., below 15 km). When sunlight consisting of ultraviolet rays, visible light and infrared, fall on the top of the atmosphere, the harmful ultraviolet rays are absorbed by the O_3 layer and so these do not reach the earth's surface. However, the visible light and infrared rays pass though the CO_2 layer and fall on the earth. The result is that the earth gets heated due to heating effect of the infrared radiations. As the earth becomes hot, it starts emitting heat (infrared rays) back to the atmosphere.

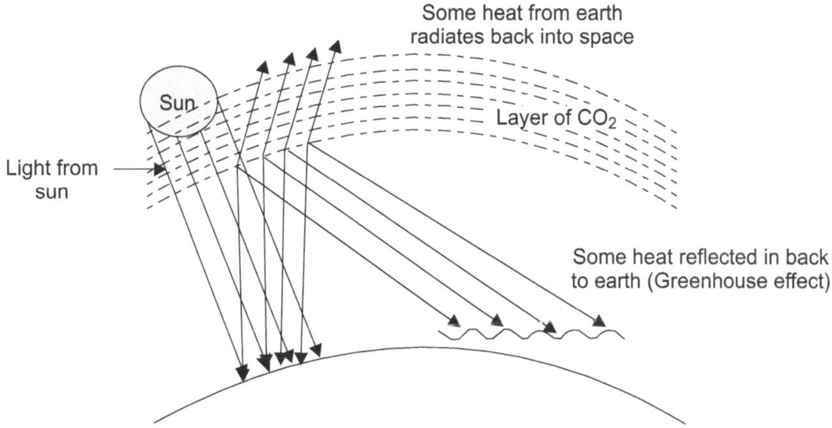

Fig. 4.3 Greenhouse effect

The emitted infrared rays are absorbed by the CO_2 layer in the atmosphere. In this way, CO_2 in the atmosphere does not allow the infrared radiation reflected by the earth's surface to go out of the atmosphere. We can say that the layer of CO_2 gas in the atmosphere traps all the infrared radiations (heat rays) coming from the earth's surface. These trapped infrared rays heat the earth's atmosphere. The heating up of the earth due to the trapping of infrared radiations (reflected from the earth's surface) by CO_2 layer in the atmosphere is called **greenhouse effect**. Thus, the temperature of the earth is raised. The rise in the temperature produced by the 'greenhouse effect' in the earth's atmosphere depends on the amount (proportion) of CO_2 gas in the atmosphere.

4.9.3 Global Warming

An increase in the average temperature of the earth's atmosphere due to green house effect which can have far reaching effects on the climate and consequently on the key life, support systems of the planet. Since the last century, human activities, mainly burning fossil fuel, has raised the atmospheric concentration of green house gases, which may intensify the green house effect and result in an increase in the average global temperature. Since global climate including rainfall, storms, wind patterns, ocean currents and sea level is intimately related to global heat flows and temperature patterns, there is a general expectation that the earth's climate may be significantly modified in the next 50 years. This would, in turn, alter the earth's delicate ecological balance.

Between the years 1878 – 1990, an increase of 0.45°C to 0.15°C has been there; during 1990 – 2025, there may be further rise by 1°C and between 2025 – 2100 there may be 3°C rise.

Besides what has been stated above, the rise in temperature will melt all the glaciers (snow-mountains), flooding the low-laying areas of the earth. An increase in global temperature is also likely to increase the incidence of infectious diseases such as malaria, dengue, sleeping sickness and yellow fever. Table 4.3 Gives some major greenhouse gases and their sources.

Table 4.3 Sources of Major Greenhouse Gases

Gas	*Sources*
Carbon dioxide, CO_2	Fossil fuel combustion (coal, oil, natural gas) Deforestation and land use changes
Methane, CH_4	Enteric fermentation in cattle and insects Biomass burning and garbage land fills Coal mines and natural gas leaks Rice paddies Swamps and tundra
Chlorofluorocarbons	Aerosols (spray propellants) CFCs Refrigeration and air-conditioning Plastic foams Industrial solvents Sterilants for medical supplies
Nitrous Oxide, N_2O	Fertilizer use Fossil fuel combustion Biomass burning Changing land use

4.9.4 Depletion of Ozone (Ozone Hole)

The ozone layer protects life on earth by absorbing the high energy ultraviolet radiations from the sun reaching the earth. These UV radiations can cause skin cancer and also damage the crop plants. In 1987, the thinning of the ozone layer was first observed over the Antarctics, which is known as **Ozone hole**. The following can cause the thinning or the destruction of the ozone layer or the depletion of the ozone layer:

(*i*) The oxides of nitrogen present in the atmosphere decompose ozone into oxygen:

$$NO + O_3 \longrightarrow NO_2 + O_2$$

$$NO_2 + O \longrightarrow NO + O_2$$

We, thus, see that the oxides of nitrogen (which are formed during thunderstorm and lightening by combination of atmospheric nitrogen and oxygen) destroy or deplete the ozone layer. It is believed that the above process destroys more than 60% of ozone in the atmosphere.

Various nuclear tests conducted in various parts of the world are also responsible for the destruction of ozone layer. Since nuclear tests produce such a high temperature that the atmospheric nitrogen and oxygen combine forming nitric oxide, which as stated above, is responsible for destroying ozone.

(*ii*) Depletion of ozone layer by chlorofluorocarbons (CFC's)–As the name indicates, chlorofluorocarbons contain carbon, chlorine and fluorine. The CFC's are non-flammable and non-toxic. They are used as refrigerants (coolants for refrigerators and air conditioners), for making polystyrene, for fire-fighting (special types of CFC's, known as halons are used as fire extinguishers; halons are bromine containing derivatives of chlorofluorocarbons), as propellant for aerosol sprays and as blowing agents for foam plastics.

Though unreactive, the CFC's are a major factor for ozone depletion. Being inert, CFC's do not easily degrade in the troposphere and so they find their way into the stratosphore which contains the ozone layer. In stratosphore, the CFC's are photochemically dissociated into reactive chlorine atoms. The latter catalyse the depletion of stratospheric ozone as shown below:

$$Cl^{\cdot} + O_3 \longrightarrow ClO^{\cdot} + O_2$$

$$ClO^{\cdot} + ClO^{\cdot} \longrightarrow (ClO)_2$$

$$(ClO)_2 + h\nu \longrightarrow Cl^{\cdot} + ClOO^{\cdot}$$

$$ClOO^{\cdot} \longrightarrow Cl^{\cdot} + O_2$$

$$Cl^{\cdot} + O_2 \longrightarrow ClO^{\cdot} + O$$

The net reaction is $2O_3 + h\nu \longrightarrow 3O_2$

It is estimated that for every reactive chlorine atom generated in the stratosphere 1,00,000 molecules of ozone are depleted. The depletion of ozone by chlorofluorocarbons can be schematically represented below in Fig. 4.4.

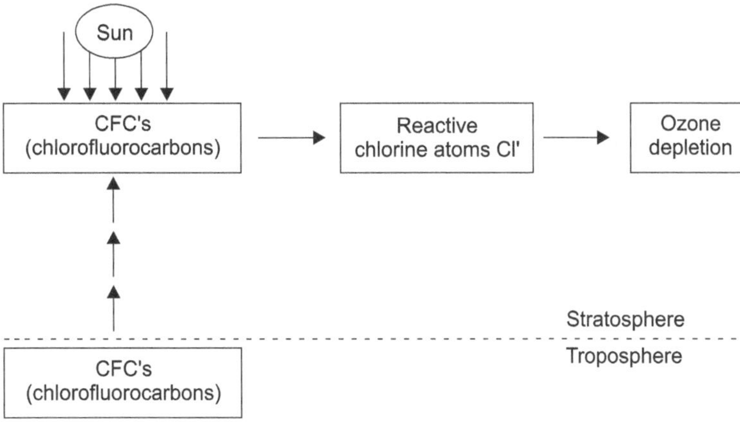

Fig. 4.4 Ozone depletion by CFC's

Concerned about the ozone hole (which permits harmful UV radiations to reach the earth), a protocol was signed at Montreal, Canada in 1987 in which representative of a number of developed countries reached an agreement to limit the production of CFC's. The target set was to reduce the use of these harmful CFC's by 50% by 1998. It was found that the ozone depletion caused by CFC's was much more alarming. In view of this, the Montreal convention was followed by a conference held in London in 1990 in which many more countries participated. The aim of the Montreal Protocol was affirmed and it was decided to speed up the phasing out of CFC's.

Prevention or Minimisation of Ozone Depletion

As already discussed ozone depletion is effected by oxides of nitrogen and CFC's. The oxides of nitrogen find their way into the atmosphere by the combination of atmospheric nitrogen and oxygen particularly during thunderstorm and lightening. There is virtually no way to prevent this mode of ozone depletion. However, using alternative substitutes can control the depletion of ozone by CFC's.

Due to inert nature and enormous stability of CFC's, they stay in the atmosphere for very long time. For example, the two most commonly used CFC's viz. CFC-11 (CCl_3F) and CFC-12 (CCl_2F_2) stay in the atmosphere for 75 and 110 years, respectively. The implies that even if the use of CFC's is stopped, the already discharged CFC's in the atmosphere will continue to deplete the ozone for a considerable length of time. In view of this, it is more or less essential to phase out the use of CFC's and their substitutes must be discovered. The substitutes should have zero potential of ozone depletion and their green house potential should also be very low. It has already been stated that CFC's are the constituents of green house gases and contribute to global warming.

The following CFC's are most likely to replace the usual CFC's. These are HCFC-123 ($CHCl_2CF_3$) and HFC–134 a (CH_2FCF_3). These have very low ozone depletion potential and their green house potentials are also quite low. However, their degradation products are toxic in nature.

It should be mentioned at this stage that certain related chemicals like carbon tetrachloride (CCl_4) and trichloroethane (CH_3CCl_3) also have ozone depleting potential. Use of these chemicals also has to be phased out. At present, there are no substitutes for these (trichloroethane is used as a solvent and carbon tetrachloride is used as a raw material for the manufacture of CFC's and pesticides and as a fire extinguisher).

Checking the Depletion of Ozone

It has been reported (Science, 1991, 254, 1191 – 1194) that the depletion of ozone could be prevented or at least minimised by injecting alkanes (e.g., ethane, propane etc) in the stratosphere (which mainly contain ozone. The alkanes will react with ozone distroying chlorine atoms (from CFC's) and immobilize item.

4.10 CONTROL OF AIR POLLUTION

The air pollutants constitute a great danger to one and all. The harmful effects of various pollutants, viz., carbon monoxide, carbon dioxide, oxides of nitrogen, oxides of sulphur, particulates have been discussed at appropriate places. Following are given the procedures for the control of some important air pollutants.

4.10.1 Oxides of Sulphur

The presence of sulphur dioxide in the atmosphere is very hazardous. It is very necessary to check its emission into the atmosphere. The possible approaches are the desulphurisation of the fuel (coal is the major source of SO_2) and removal of SO_2 from flue gases.

(*i*) Desulphurisation of Coal: Coal contains sulphur either in inorganic from (as iron pyrites) or inorganically bound form, in approximately equal ratio. The inorganic sulphur is removed by subjecting grounded coal to hydraulic washing. Iron pyrites, being heavier settle down while the sulphur free coal floats on the surface and is collected. The organically bound sulphur is removed by passing hydrogen gas over the fuel in presence of catalyst (mixed cobalt molybdenum oxide). By this process, the organically bound sulphur is converted into hydrogen sulphide, which is absorbed in a solution of diethanolamine, subsequently H_2S is oxidised to sulphur.

$$H_2S + [O] \longrightarrow H_2O + S$$

The organic sulphur can also be removed from coal by a microbial method using a strain designated as IGT-57. However, this method does not remove inorganic sulphur and is too slow to be of utility.

(*ii*) Desulphurisation of Flue Gases: In this procedure, powdered coal and powdered lime stone are heated, when the following reactions take place.

$$CaCO_3 \xrightarrow{\Delta} CaO + CO_2$$

$$CaO + SO_2 \longrightarrow CaSO_3$$

$$CaO + SO_2 + O \longrightarrow CaSO_4$$

The formed calcium sulphite and calcium sulphate are removed by scrubbing with a slurry of lime water obtained by hydrolysis of lime, $CaO + H_2O \rightarrow Ca(OH)_2$.

The method is economical but the disposal of waste $CaSO_4$ in large amount creates problems.

Alternatively, SO_2 in flue gases is reduced by mixing it with carbon monoxide (from another polluting unit) over a catalyst.

$$2CO + SO_2 \xrightarrow{\text{Catalyst}} 2CO_2 + S$$

A number of catalysts have been used. The disadvantages of the catalytic method are the cost and deactivation of the catalyst.

Alternatively, SO_2 in the exhaust gases is removed by using scrubbers. The scrubbers force the gases through a slurry of calcium carbonate, or magnesium hydroxide. The SO_2 dissolves in water and reacts with $CaCO_3$ or $Mg(OH)_2$ to form a salt. If $Mg(OH)_2$ is used, the $MgSO_3$ that is formed can be isolated and heated to regenerate SO_2, which can be collected and used as a raw material for other industrial processes.

$$SO_2 + Mg(OH)_2 \longrightarrow MgSO_3 + H_2O \text{ (removal of } SO_2\text{)}$$

$$MgSO_3 \xrightarrow{\Delta} SO_2 + MgO \text{ (regeneration of } SO_2\text{)}$$

Fig. 4.5 Below gives a schematic diagram of a scrubber.

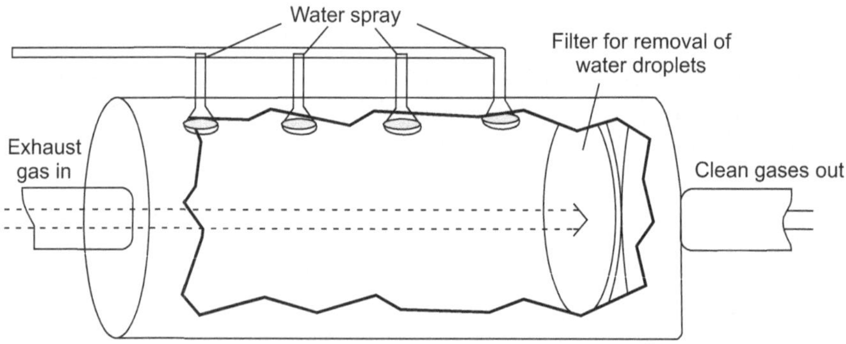

Fig. 4.5 Scrubber (schematic diagram)

In both the above methods, the oxides of nitrogen which are invariably present in flue gases are not removed. Both oxides of nitrogen and sulphur can be removed using a single technology as described below.

4.10.2 Oxides of Nitrogen

It has already been stated that the important sources of oxides of nitrogen in the atmosphere are fossil fuel based power plants and automobile exhaust – both of which require control procedures.

1. Checking the emission of oxides of nitrogen (and simultaneous checking emission of oxides of sulphur) from coal. This is affected by the so called coal conversion technology.

The procedure involves heating powdered coal with sodium hydroxide solution under moderate pressure. This removes sulphur along with many trace elements from coal. The coal obtained is free of sulphur and will not produce oxides of sulphur on combustion. The sulphur free coal is reacted with steam to produce water gas (a mixture of carbon monoxide and hydrogen; coal + steam $\rightarrow$ $CO + H_2$). The water gas is sprayed with water and burned. The presence of water lowers the combustion temperature and so nitrogen and oxygen do not combine and form oxides of nitrogen.

Alternatively, **fluidized-bed combustion** technique is used. In this method, a bed of ash or sand is placed on a grate at the bottom of a boiler. Air is blown upwards through the great and simultaneously crushed solid fuel is fed into the boiler. The fuel heats up the particles of ash or sand. The heat is picked up by the coils in which water is fed at the bottom and steam comes out from the top (see fig.4.6). Subsequently, limestone is added to the mixture in the boiler. Limestone combines with the evolved SO_2 from the fuel.

$$2CaCO_3 \xrightarrow{\Delta} CaO + CO_2$$

$$2\,CaO + 2\,SO_2 + O_2 \longrightarrow 2\,CaSO_4$$

The system operates at 800–1000°C. At this low temperature, oxides of nitrogen are not formed. So, the flue gases are free from both the oxides of sulphur and nitrogen.

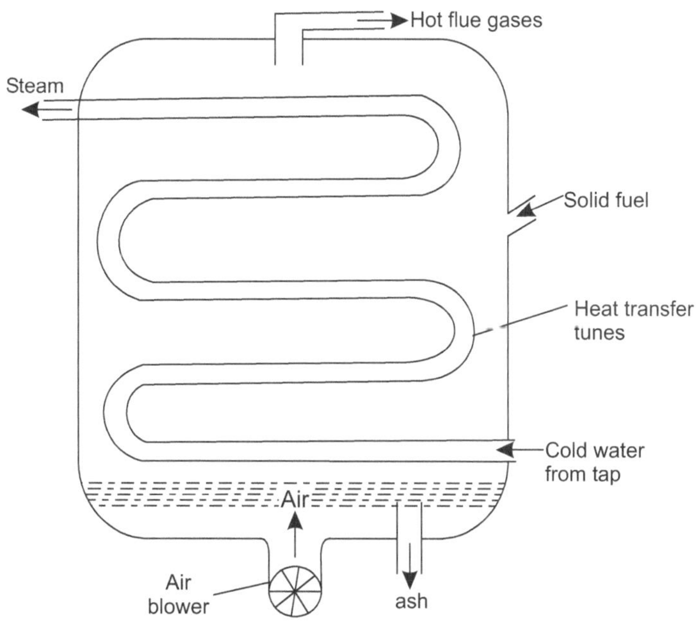

Fig. 4.6 Fluidized bed combustion

4.10.3 Checking the Emission of Nitrogen (and Simultaneously Carbon Monoxide and Unburnt Hydrocarbons) from Automobile Exhaust

The automobile exhaust has become a major source of atmospheric pollution (oxides of nitrogen and carbon monoxide) since the number of automobiles has increased enormously. The automobiles exhaust pollutants are controlled by installing catalytic converters in the exhaust system, so that the exhaust gases are first passed through a catalytic converter before being released into the atmosphere as harmless gases like CO_2, H_2O and N_2. The only limitation of this procedure is that the gasoline or petrol should be unleaded since lead poisons the catalyst. It may be appropriate to state that anti-knocking compounds like lead tetraethyl are now not added to the petrol. Instead, the composition of the petrol is so adjusted (by the addition of some branched chain hydrocarbons) so as to have the desired octane rating. The catalytic converter contains a mixture of plantinum, palladium and rhodium supported on a honeycomb bed (made of ceramic or metal). Various reactions taking place in the catalytic converter are:

$$2\,CO + O \xrightarrow{\text{Catalyst}} 2\,CO_2 \quad (i)$$

$$\underset{\text{Hydrocarbon}}{HC} + O_2 \xrightarrow{\text{Catalyst}} CO_2 + H_2O \quad (ii)$$

$$2NO + 2CO \xrightarrow{\text{Catalyst}} N_2 + 2CO_2 \quad (iii)$$

$$2\,NO + 2\,CO \xrightarrow{\text{Catalyst}} N_2O + CO_2 \quad (iv)$$

$$2\,N_2O \xrightarrow{\text{Catalyst}} 2\,N_2 + O_2 \quad (v)$$

$$NO + HC \xrightarrow{\text{Catalyst}} N_2 + CO_2 + H_2O \quad (vi)$$

As seen, in the catalytic converter

(*a*) CO gets converted into CO_2 (eq. (*i*))

(*b*) Hydrocarbons are oxidised into CO_2 and H_2O (eq. (*ii*))

(*c*) Nitric oxide reacts with CO to give N_2 and CO_2 (eq. (*iii*))

(*d*) Nitric oxide may also be partially reduced to N_2O (eq. (*iv*)) and the formed N_2O decomposes into N_2 and O_2 (eq. (*v*))

(*e*) Nitric oxide and hydrocarbons are converted into N_2, CO_2 and H_2O (eq. (*vi*))

Carbon Monoxide

Carbon monoxide is the most harmful pollutant (being poisonous in nature) in the atmosphere. Its main sources are incomplete combustion of all carbon containing fuels particularly in automobile engines and industrial furnaces and cigarette smoking. It is removed from the source in the following way.

(*i*) From automobile exhaust gases, it is removed (along with oxides of nitrogen and unburnt hydrocarbons) by installing catalytic converter in the exhaust system.

(*ii*) Accumulation of carbon monoxide in any space (like in submarines and space flights) is removed by providing catalytic burner to oxidise CO to CO_2.

(*iii*) Cigarette smoking should be done only in specific area with full ventilation. However, it is not possible to control CO from the cigarette smoke. The best way is to quit smoking.

Control of pollution from polymers and plastics (page 94), asbestos (page 125–128) are given at appropriate places.

4.10.4 Particulates

The common sources of particulate emission into the atmosphere are the powerhouses operating on coal. The smoke that comes out of the chimneys contains fine particles of carbon which remain suspended in air. These particulates are removed by installation of an electrostatic precipitator in the chimney. By this process when a high electrostatic potential is applied across the chimney, the carbon particles coming out of the chimney are attracted by charged electrodes. This results in the formation of soot which gets deposited in the chimney and is cleaned from time to time to remove soot. The particulates coming in the exhaust of automobiles are removed by catalytic converter.

A schematic diagram of an electrostatic precipitator is shown in Fig. 4.7. As shown, the exhaust gases from power houses and other industrial houses (which used coal or other fuels as a source of energy) are passed between two charged vertical plates or electrodes. The particles pick up electrons supplied by the negative electrodes and move to the positive one. Thus, the liquid particles (if present as in aerosols, which are small particles of liquid and solids dispersed in smoke) accumulate and flow to the bottom of the collector. The solid particles drop off with agitation.

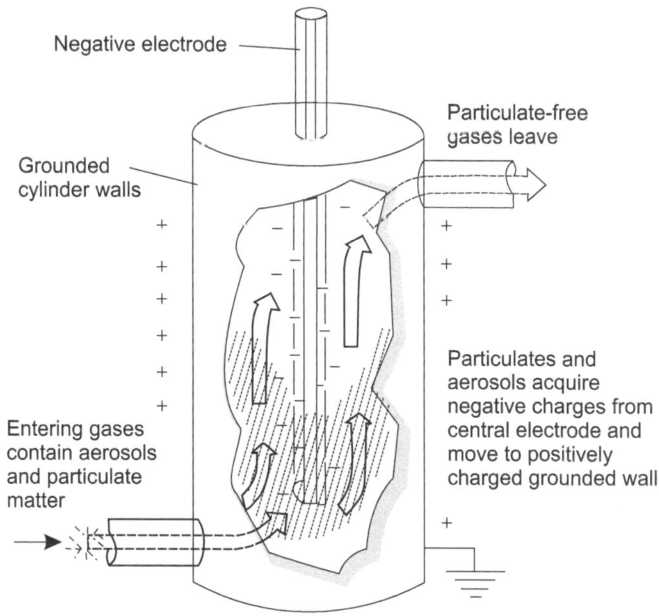

Fig. 4.7 An electrostatic precipitator

In some of the industries, the filters (called the industrial filters) are used to remove particulates. These filters work like the filters used in air conditioners or vacuum cleaner filters. The exhaust gases are passed through them, leaving particulates and aerosols behind the filter. A typical industrial filter is shown in the Fig. 4.8.

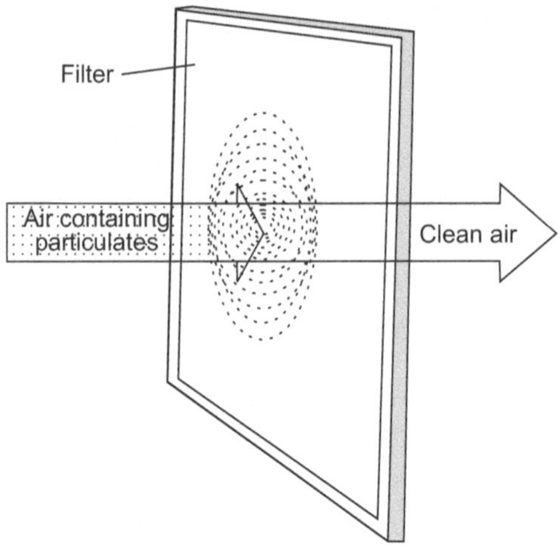

Fig. 4.8 An industrial filter

Use of Alternative Source of Energy

Following are given some of the alternative source of energy:

(*i*) Use of hydroelectric power (if feasible) in place of power generation by coal.

(*ii*) Use of nuclear power. This has its own problems of pollution, disposal of the waste and nuclear accidents.

(*iii*) Use of alternative energy sources for automobiles. Of these natural gas (CNG) is particularly appealing because of least potential for air pollution. In fact, in Delhi, it is now mandatory for taxis and buses to use CNG.

(*iv*) In some of the advanced countries hydrogen is used as fuel for automobiles. In India hydrogen is mixed with CNG to an extent of 20–25% for automobiles.

Electric cars powered by batteries may seem to be completely pollution free alternative. However, their use might only shift environmental concern from one source or location to another. The electricity requirement for charging all the transportation batteries would have to be met by existing or new electric power generators which by itself may be of environmental concern.

4.11 ANALYTICAL METHODS FOR MEASURING AIR POLLUTION

It is very important to find the quality of air we breathe. For this, analytical methods have been developed for measuring the extent of air pollution. In 1971, the United States Environmented Protection Agency (USEPA) set national quality standards as given in Table 4.4

Table 4.4 USEPA Air quality Standards for Primary Pollutants

Pollutant	*Tolerance (ppm)*	*Levels μg/m³*	*Relative toxicity*
Carbon monoxide (CO)	9.0 (not to be exceeded more than once a year for 8-hour period). 35.0 (not to be exceeded more than once a year for 1 hour period)	10,000 40,000	1.0
Hydrocarbons	–	19,3000	2.07
Oxides of sulphur (SO_x)	0.50	1,430	28.00
Oxides of nitrogen (NO_x)	0.25	514	77.8
Particulates	–	375	106.7

Source: National Air Quality standard, Environmental Science and Technology, 1971,5,503

The qualitative relative toxicity values as given in Table 4.4 is more important than their tonnage wise amounts, since it does not take into account the possibility of one pollutant being more harmful than the other. For example, particulates are the most toxic pollutants, but tonnage wise carbon monoxide has been found to be most toxic. For the purpose of measurements, air pollutants are classified into five groups:

I. Materials like CO, SO_x, NO_x, hydrocarbons, particulates etc. for which standard have been set (Table 4.4).

II. Pollutants, which are hazardous to human health, but for which ambient air quality standards are not specified. Examples include asbestos, Ne and Hg.

III. Pollutants which are visible emissions. Examples include pollutans from coal – cleaning plants, cotton gins, lime plants and paper mills. Such pollutants require frequent monitoring.

IV. Pollutants from mobile sources, eg, emissions from automobiles etc.

V. Elements and compounds which require control for public health. Examples include, As, Cd, Ni, Cr, V, Li, Cu, Zn, Ba, Sn, F^-, Cl_2, HCl, H_2S, Se, B, P, PAH, polychlorinated biphenyls, fine particles, pesticides, reactive organic compounds and radionuclides.

The **units** used for various pollutants depend on the nature of the pollutant. Some important units are given below

Gaseous and vapours, μg/m³ (also ppm by volume)

Weight of Particulate matter, μg/m³

Count of particulate matter, number per cubic meter

Emission and sampling rate, m³/min

Pressure, mm Hg

Temperature, degrees centigrade

Visibility, Km.

4.11.1 Sampling

It is necessary to obtain a good sample for analysis. For this various relevant factors should be taken into account. Generally speaking a sample of about 10 cm^3 (cubic meters) is required and the optimum sampling limit is 3 hour.

Following are given some of the important and frequently used procedures for sampling particulate matter and gaseous pollutants.

Particulates

The particulate matter is collected by filtration technique. For metals and anions, cellulose filter paper is used. Organic components are collected by using glass fibre for filtration. Silica felts are used for trace inorganic species and organic compounds. A typical instument for filtration is shown in Fig.4.9.

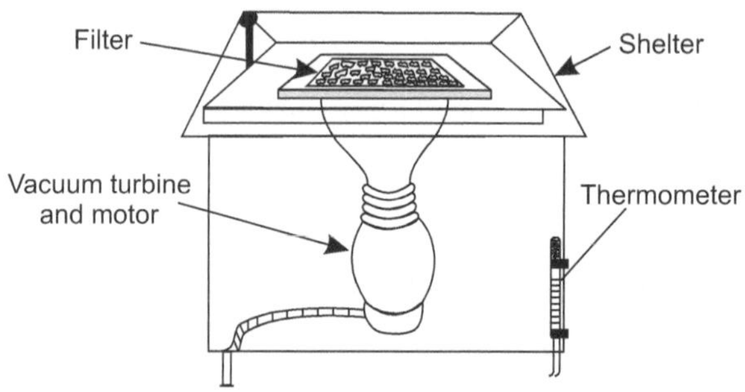

Fig. 4.9 Sample collection of particulates by filtration

Electrostatic Samplers

Also known as electrostatic precipitators, these are routinely used for the control of particulate matter. These are made of a number of collector relates (rectangular in shape) and are spaced apart with the help of a wire discharge electrode (Fig. 4.10). A strong electric fields applied to the wires and plates. The flue gases like carbon dioxide, oxides of nitrogen and sulphur passes between the electrodes, these are ionized and the formed ions charge the particle, which are attracted by the electrostatic field force to the oppositely charged electrodes. The particulate matter gets accumulated on the plates and get collected via the hoppers.

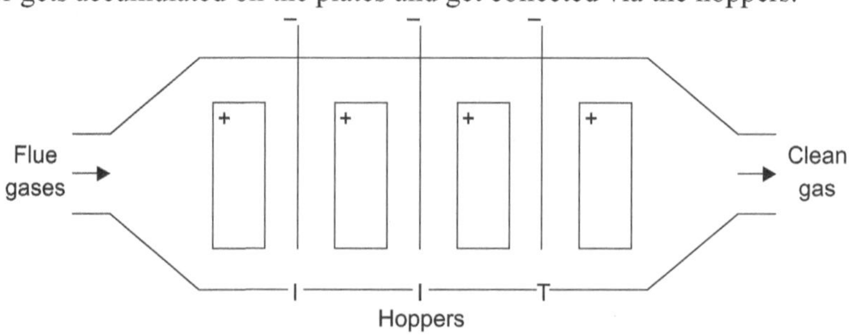

Fig. 4.10 Simple representation of an electrostatic precipitators

Electrostatic samplers can collect about 99% or more of particulate matter and can also be used for the reduction of air-borne lead emissions.

Thermal precipitators

Thermal precipitators are based on the principle that hot surface repel particles suspended in air and repelled particles are then collected on cold surface. The suspended particles thus move towards lower temperature regions exposed to high temperature gradients. This procedure is useful for using in coal fired electric generating Power Plants and also suitable for collection of aerosol particles of 0.001 μ size.

Gaseous Pollutants

A commonly used method for collection of gaseous pollutant is **absorption** is a suitable solvent. The method consist in bubbling the gas through the solvent by using small bubble size and increased residence time. Pure water is good for collecting some gaseous pollutants like HF. Alkaline solutions are required for absorbing acidic gases and acidic solutions for collecting alkaline gases and oils for hydrocarbon. A typical apparatus for absorbing gaseous pollutants is shown in Fig. 4.11.

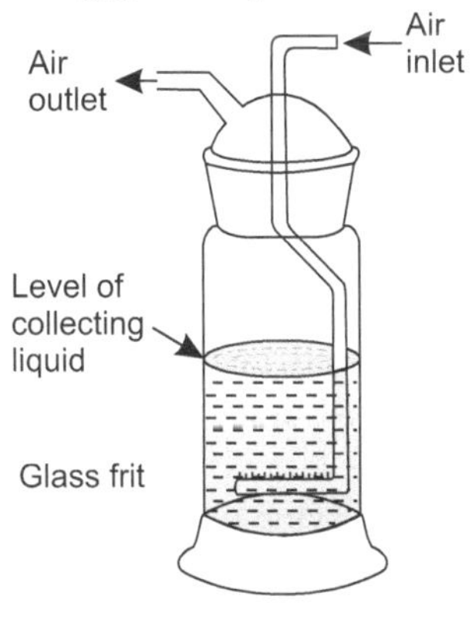

Fig. 4.11 A typical apparatus for sampling air pollutions

Adsorption can also be used for gaseous pollutants. Generally charcoal is used as an efficient adsorbant. Besides charcoal, Al_2O_3, silica gel and molecular sieves can also be used as adsorbant. The fig below (Fig 4.12) shows a usual design.

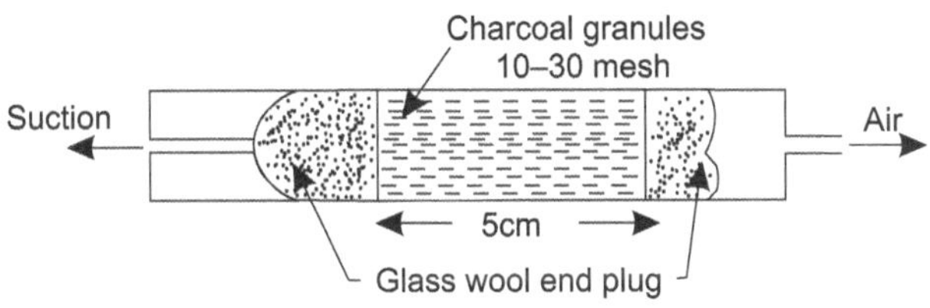

Fig. 4.12 Charcoal absorption tube

After the adsorption, the sample can be removed for analysis by heating the column and flushing it with gas or by applying a vacuum.

4.11.2 Analysis of Gaseous Pollutants

Determination of the concentration of carbon monoxide, oxide of sulphur and oxides of nitrogen has already been described in pages 115, 118 and 120 respectively.

Analysis of Hydrogen Sulphide

Hydrogen sulphide normally coexists in the atmosphere with SO_2. Hydrogen sulphide is produced by the reduction of sulphur and organosulphur compounds by the bacterium Desulphovibrio desulphuricans and associated with methane thiol (CH_3SH), dimethyl sulphide ($CH_3.S.CH_3$), etc. The TLV of H_2S is 10 ppm (TLV is the concentration of an individual chemical is air to which workers may be exposed day after day without any adverse effect). As H_2S is readily oxidised is air, it is trapped as the insoluble metal sulphide (CdS, ZnS, HgS or Ag_2S) in the absorption reagent. Following procedure is followed.

Pump 100–500 L of air at 10–15 L min^{-1} through a wash bottle into a 50 mL 1 per cent solution of $CdSO_4.8H_2O$ and 50 mL of 0.01M NaOH solution taken in a standard gas wash bottle with filled glass disperser. Add 2 ML. of 1 M $NaClO_4$ and shake well. This is followed by the addition of 1 mL of 2 per cent W/V N, N-diethyl *p*-phenylenediamine in 50 per cent W/V H_2SO_4 and 1 mL of a filtered 10 per cent W/V solution of iron (III) ammonium sulphate dissolved in 1 M H_2SO_4. The solution is transferred to a separatory funnel, extracted twice with 10 mL $CHCl_3$. The combined extracts made up to 25 mL and absorbance of the formed ethylene blue dye measured in a 1 cm cell at 670 nm. For a calibration curve, use stock solution of Na_2S (1 – 10 μg range) prepared by washing the crystals of Na_2S with ethanol and drying on filter paper.

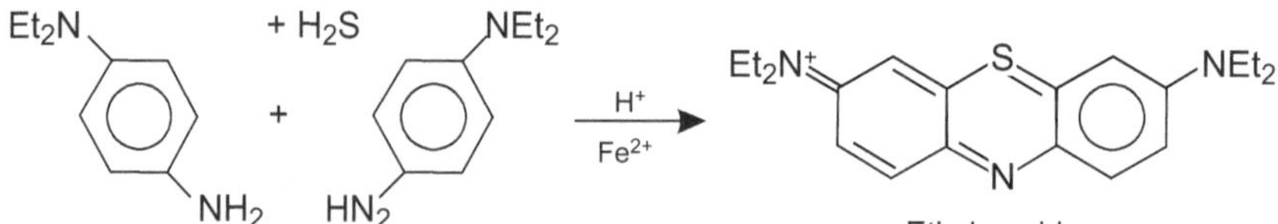

Ethylene blue

Alternatively, UV molecular absorption method can also be used for determination of H_2S. In this procedure H_2S is collected in $(CH_3COO)_2Zn$, the sample made acidic and the liberated H_2S gas swept in a steam of N_2 down a 20-cm-path-length optical cell, positioned where flame would normally be in a atomic absorbance spectrophotometer. The absorbance is continuously recorded. A Mg hollow cathode lamp is used with the spectrophotometer tuned to 202-mm line.

Analysis of Hydrocarbons

Hydrocarbons are collected is an absorption column packed with a very porous styrene—divinylbenzene polymer (as shown in Fig. 4.12). Air is sampled at the rate of 4 L/mim. A total of 10L sample of air is required. The sample collector is subsequently heated to desorb the sample for gas chromatographic analysis. As little as 0.1 mg of hydrocarbon/m^3 of air can be estimated using a flame ionization detector for gas chromatograph.

For low levels of hydrocarbons, the gas chromatograph is used in conjuction with a mass spectrometer (GC/MS). This procedure is extremely helpful for the identification of hydrocarbon mixture.

The hydrocarbons in automobile exhaust gases is determined as follows.

Place a bag made of polyethylene or polyester inside a 1 litre Jar. Fit up the apparatus as shown in Fig 4.13. First below air into the space between the bag and 1 litre Jar through the tube (1). In this way the contents of the bag are expelled through tap and tube (2). Next suck through tube (1), connecting (2) with exhaust gas. In this way the bag gets filled with exhaust gas.

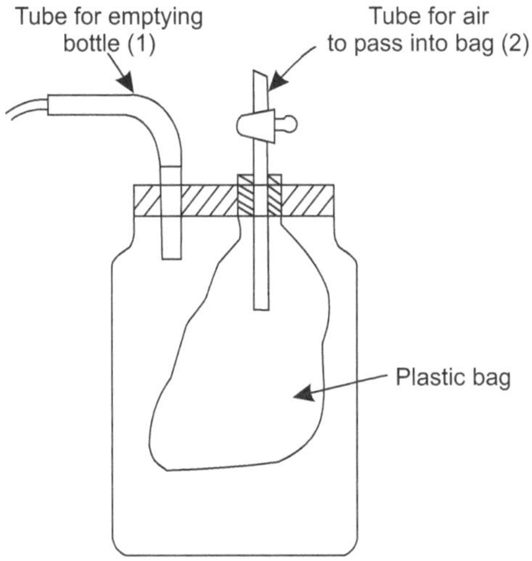

Fig. 4.13 Sampling of exhaust gas

Finally, use a 1 mL loop on the sampling valve, a single carbowax 1540 column at 90°, with 10 psi of N_2 carrier gas and a flame ionization detector (FID). Inject 10 μL portions of the sample of exhaust gas (collected in the bag) into the system by a syringe through the injection port. Different peaks in the GC represent different hydrocarbons, which are identified by using known samples and noting the retention times (for identification). For estimating the amounts, the peak—heights (for a 10μL portions of solutions of individual compounds) are compared with the peak height in the exhaust gas sample.

For comparing the hydrocarbons in the exhaust gas it is helpful to find different hydrocarbons in the petrol itself. For this 0.1 mL of petrol is diluted to 10 mL with CS_2. Mix well and inject a 5μl portion. Identity the peaks where possible and make quantitative estimates. The petrol normally contains benzene and toluene (5% and 10% V/V respectively) along with other hydrocarbons like ethylbenzene, *o*, *m*, *p*-xylenes, mesitylene and 1, 2, 4-trimethyl benzene.

Analysis of Particulate Matter

Amount of Particulate matter is determined by using electrostatic precipitators (see Fig. 4.10). Initially the weight of the total instrument is determined. Subsequants air is drawn at a rate of at least 1.70 cm^3 min^{-1} as measured by a rotameter. After sampling for 24 hr, the instrument is weighed again. The difference gives the weight of the particulate matter. The particulate matter may be extracted with a suitable srevent or incinerated or digested by oxidizing mixture before elemental analysis

4.11.3 Instrumental Methods for Measuring Air Pollution

Instrumental methods are now been used because of extreme rapidity with which analysis can be carried out. These methods give accurate results and are very efficient. These methods find application for continuous monitoring air pollution. Following are given some of the instrumental methods which are commonly used.

Spectrophotometric Method

The spectroscopic methods are faster, accurate and require much less sample,

Ultraviolet Spectroscopy

Ultraviolet spectroscopy is an important tool, which is helpful for the identification of organic compounds. However, it cannot be used to show the presence of individual functional groups (as is the case with infrared spectroscopy, discussed subsequently), but it gives information about the presence and extent of multiple bonds and conjugation in organic molecules. This technique is routinely used to show relationship between functional groups, particularly conjugation between two or more carbon-carbon double (or triple bonds); between carbon-oxygen double bond and the aromatic moiety. It also indicates the number and location of the substituents attached to the conjugated system.

The region of primary interest to an organic chemist is between 200 to 380 nm. This region is known as the near UV region. Most of the unsaturated compounds absorb in this region. The region 380 to 80 nm is called visible region, in which all coloured compounds absorb. The UV portion of the electromagnetic spectrums is shown to fig. 4.14

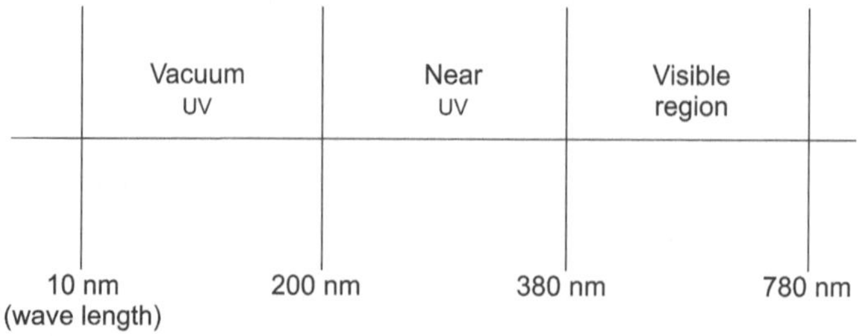

Fig. 4.14 The UV portion of the electromagnetic spectrums

On subjecting a compound with multiple bands to electromagnetic radiation in the UV/visible region, a portion of the radiation is absorbed by the compound. The amount of absorption depends in the structure of the compound and the wave length of radiation. The ultraviolet spectrum is a plot of wavelength (or frequency) of absorption VS the absorption intensity (absorbance or transmittance). Thus, the uv spectrum records the wavelength of an absorption maximum which is represented by λ max in nanometers (1nm = 10^{-9}m). The intensity of absorption is given by the Beer-Lambert Law.

$$\log (Io/I) = \varepsilon \text{ c.l.}$$

$$\varepsilon = A/\text{c.l.}$$

Where Io is the intensity of the incident hight; I is the intensity of the transmitted radiation. A is absorbance (optical density); ε is molar absorptivity (molar extinction coefficient); c is the concentration of the solution (g. mol/L) and *l* is the length of the cell containing the solution (in cm).

The plot of the curve of molar absorptivity (ε) or logarthm of molar absorptivity (log ε) versus wave length (λ) constituents the absorption band (Fig. 4.15).

The molar absorptivity ε is constant for an organic compound at a given wave length, and is expressed as ε_{max}, i.e., molar absorptivity at as absorption maximum.

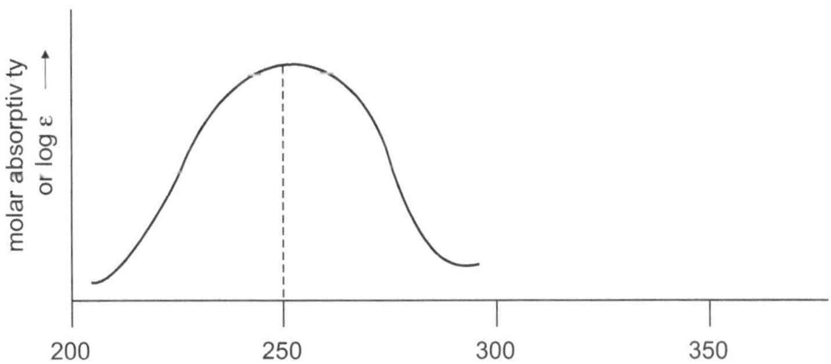

Fig. 4.15 Representation it a UV absorption band

Organic substances give rise to several transitions involving both bonding as well as non banding electrons. Electronic transitions in organic molecules usually involve transitions of sigma (σ) electrons, *n*-electrons, and π-electrons. σ-electrons are located in σ-bonds. For example, a single valence bond is between two carbons as found in saturated hydrocarbons. The σ-electrons are tightly held and energy of uv and visible region cannot sufficiently overcome this attraction. On the other hand, *n*-electrons are less firmly held than σ-electrons are non-bonding electrons and are found is atoms such as N, O, S and halogen. In case of π-electrons, the energy of UV and visible region is sufficient to cause the excitation process is take place. The *n*-electrons may undergo two type of transistions.

$$n \longrightarrow \pi^* \qquad\qquad n \longrightarrow s^*$$

An unsaturated bond contains 4 electrons (2π and 2σ-electrons). Of these π-electrons are easiest to exite. Generally the transition of a π-electron results is the absorption in the ultraviolet or visible region. Aromatic compounds have characteristic UV absorption spectra due to their π-electrons.

For inorganic ions and molecules, the transitions may occur between the two electronic states of the individual atoms or there may be partial electron transfer from ligand to the central ion resulting is a charge transfer absorption spectrum. Absorption spectrum of the substances may be a result of contribution of the electronic transition and the electron transfer process.

Infrared spectroscopy

The infrared radiations do not cause excitation of electrons (as in the case of UV spectroscopy) due to less energy associated with them. The radiations, however, cause vibration of atoms within the molecules which are quantised and absorb infrared radiations of a particular energy.

In case of monatomic particles (such as mercury or sodium), UV and visible radiation have sufficient energy to cause transition of the outermost or bonding electrons only. However, absorption by polyatomic molecule, the total energy of a molecule is given by

$$E = E \text{ electronic } + E \text{ rotational } + E \text{ vibrational.}$$

Where E is the electronic energy of the molecule, while E vibrational is the energy of the molecule as a whole due to interatomic vibrations of various energies. E rotational is the energy associated with the rotation of the molecule around its center of gravity. For vibrational, require less energetic 1R radiations is the range 1 – 15μ. Rotational changes require even less energy is the 10 – 100μ range.

It is obvious that is case of an atom, there can be no vibration or rotation as in the case of a molecule. This implies that with IR region (200 – 4000 m^{-1}) an atom does not have vibrational or rotational energy levels. However, diatomic molecules with permanent dipole, *i.e.*, heteronuclear diatomic molecules and polyatomic molecules with and without permanent dipole moments. The vibrational rotational spectras are not exhibited by homonuclear molecules, as they have no permanent dipoles. The presence of organic compounds is the air can be identified by using IR spectra. This is because every pure substance has its own characteristic IR spectra.

A simplified diagram of the various parts of the IR spectrophotometer is shown in Fig 4.16.

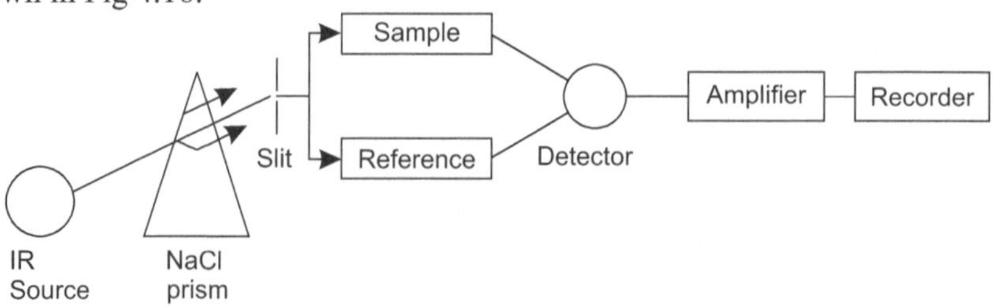

Fig. 4.16 The IR spectrometer

A continuous beam of IR radiation produced by a heated filament is separated into radiation of different wave lengths by passing through a prism made from sodium chloride. Rotation of the prism permits the different wavelengths to pass through the focussing slit to produced IR radiation of a single wavelength. The beam then splits into two beams of equal intensity, one passing through the sample and the other through reference. Subsequently the beam strikes an IR detector, which converts the transmittance into an electric signal that is amplified and sent to the recorder that traces IR spectrum.

Non-Dispersive Infrared Spectrometry

This technique has been introduced for the analysis of CO (see page 115).

In non-dispersive IR spectrophotometry, the IR radiation from a source is not dispersed according to wavelength by a prism as in a standard IR spectrometer. This technique has been used to monitor gas streams and in air pollution studies. It depends mainly on selective filtering for specificity. The radiation from an 1R source is 'chopped' by a rotating device so that it alternately passes through a sample cell and a reference cell (Fig. 4.17). Both the beams are incident on a detector which is filled with CO gas. The detector is divided with two compartments by a flexiable draphragm. The relative amounts of IR radiations absorbed by CO in the two sections of the detector are determined by the CO level in the sample. The difference in the amount of IR radiation absorbed is the two compartments give rise to temperature difference so that the diaphragm even to a slight extent can be measured with relative accuracy of 5 per cent.

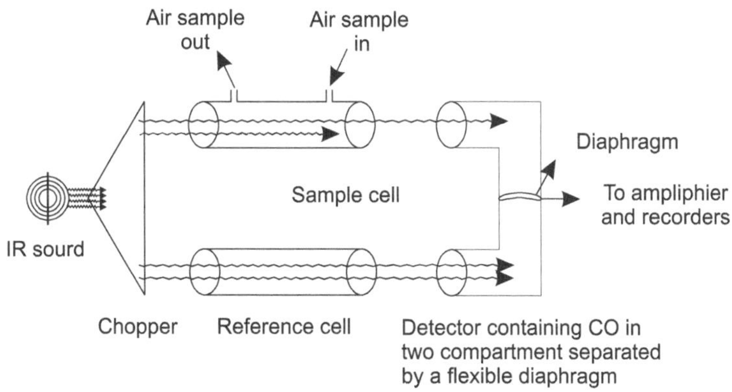

Fig. 4.17 Non-dispersive IR analyser for monitoring CO

Fourier Transform Infrared (FTIR) Spectroscopy

Infrared (IR) spectroscopy is used commonly for the identification and characterization of inorganic and organic compounds. The usual infrared (IR) spectrometer are slow and take more time. For analysis of identification in a number of constituents for example as in gas chromatography, high speed instruments are required. The best way to achieve accurate results in a very short time is to use Fourier Transform Infrared (FTIR) spectroscopy. Such instruments use the principle of Fourier transform. In this technique, the infrared radiation from a source is split into two beams (which are perpendicular to each other and are of equal intensity) using a beam splitter. Both the beams are reflected by using plane mirrors.

The FTIR spectrometer is coupled to a mass spectrometer and then to a gas chromatography (GC/FTIR/MS). This procedure is rapid for the identification of products from a GC analyser. A schematic representation of a FTIR spectrometer is given in Fig. 4.18.

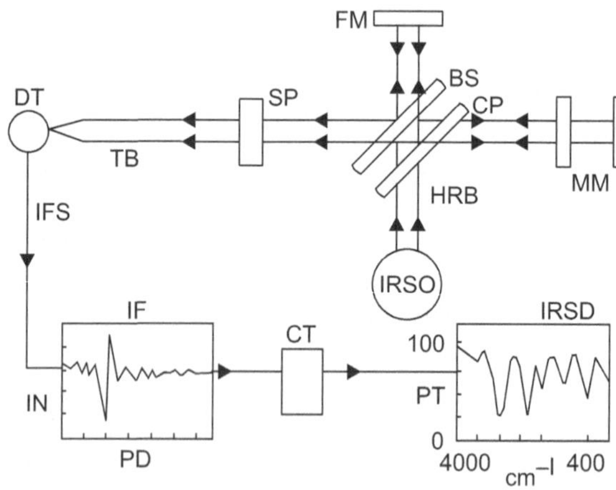

Fig. 4.18 Schematic diagram of FTIR spectrometer (IRSO=IR source and optics; HRB=incident IR beam; MM=Movable mirror; CP=compensator; BS=beam splitter; FM=fixed mirror; SP=sample; TB=transmitted beam; DT=detector; IFS=interferogram signal; IF=intensity; PD=path difference; CT=computer, PT=% transmittance; IRSD=IR special display) [G.Thygrajan; "Fourier transform infrared spectrocopy" Indian society of analytical scientists newsletter, 2, 7–12, september 1984]

Emission Spectroscopy

In this spectroscopic procedure, atoms or molecules are excited by radiations produced by thermal or electrical sources. These return to their lower states, the characteristic radiation is emitted in the form of discrete wavelengths of lines, called spectral lines. The analysis of the emitted light is carried out using a spectrophotometer, which sorts out the records of spectral lines according to their wavelengths. These lines are then identified using available table and charts. The intensity of a line emitted by an atom or molecule is a function of the concentration of the atom or molecule is the emitting source. Emission spectroscopy is useful is the identification of elements present in particulate matter e.g., Na, K, Ca, Be, Pb, Cd, Sn, etc.,

Atomic Absorption Spectroscopy (AAS)

The principle of atomic absorption spectroscopy (AAS) is the conversion of the sample to be analysed into atomic vapour state followed by measurement of the absorption at a selected wavelength (characteristic of each element). The concentration of the sample is proportional to the measured absorption. Thus, in AAS, the element is reduced to the elemental state, which is then vapourised and injected in the radiation beam from a suitable source. The various components of atomic absorption spectophotometer are depicted in Fig. 4.19

In atomic absorption spectrometer, a hollow cathode lamp is the source of radiation. Such lamps are now commonly available for all elements that can be analysed. The cathode tube is filled with an inert gas like argon under pressure. Other components include an anode and a hallow cathode. The inside of the cathode is coated with a thin film of the metal to be analysed. On the application of high potential across its electrodes, argon gas inside the cathode gets converted with Ar^+ ions due to electric current of several milliamperes which flows through the lamp. The Ar^+ ions produced attack on the surface of the metal coating of the cathode resulting in sputtering to the metal ions. The radiation emitted from metal atoms are made to pass through a flame (in the burner) into which the sample to be analysed is injected by an atomizer. In the flame, the metallic component decompose resulting in the formation of a cloy to atoms (which are now in the elemental state). The radiation in flame is partly absorbed by the cloy of atoms; this decreases the radiant energy in the radiation. Finally in resultant radiant energy passes to a monochromator and finally to the detector.

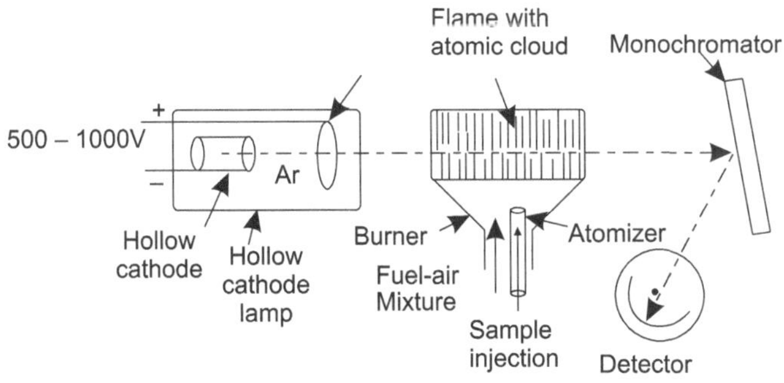

Fig. 4.19 Atomic absorption spectrophotometer showing its various components

The hollow cathode tubes are now available for all elements that can be determined by an atomic absorption spectrometer.

A modification of the AAS is the replacement it an atomizer and the sample holder, for some elements, by a graphite furnace. The detection limits with a graphite furnace are 1000 times better than those from a conventional flame. The table 4.5 gives the comparative limits of various metallic pollutants as determined by Flame AAS and graphit furnace AAS.

Table 4.5 Comparative Limits of Pollutants (ppp or μg/L) by Falme AAS and Graphite Furnace AAS. [L. Herbert, S.B Khan, R.G. Smithian, Scheicher, American Laboratory 65-79, Aug. 1979]

Element	Flame AAS	Graphite furnace AAS
Al	20	0.004
As	100	0.06
B	1000	–
Cd	1	0.008
Co	5	0.03
Cr	3	0.005
Cu	2	0.008
Fe	5	0.003
Mn	3	0.004
Mo	10	0.06
Ni	8	0.02
P	10	3.0
Pb	10	0.03
Pt	50	0.45
Se	100	0.10
Si	60	0.10
Ti	50	0.30
U	7000	–
V	20	0.15
Zn	0.6	0.0001

Gas chromatography

It is well-known that chromatography is a separation process depending on the differential distribution of the component of a mixture between a mobile phase (which can be liquid or gas) and a stationary phase (which can be liquid or solid). Depending on the state of the two phase, the chromatography can be of following type.

Liquid-solid chromatography (like columns chromatography in which the mobile phase is liquid and stationary phase is an solid adsorbent).

Gas-liquid chromatography (in which the mobile phase is gas and stationary phase is liquid).

Gas-solid chromatography (in which the mobile phase is gas and stationary phase is solid).

We are concerned in this section with gas chromatography (in which case the mobile phase is gas and the stationary phase is either liquid or solid).

Gas chromatography is a very handy and convenient method for the analysis of volatile liquids, vapours, gases and low levels of pesticides in the environment. It is based on the principle that when a mixture of gases is passed through a column containing an absorbent solid phase with the help of a carrier gas (Ar, He, or N_2), each volatile compound is partitioned between the carrier gas and the solid absorbent. Each component, depending in its retention time in the column, comes out of the column at different times and is detected by a suitable detector. This procedure permits both qualitative and quantitative analysis. A flow diagram of a gas chromatograph is shown in Fig. 4.20.

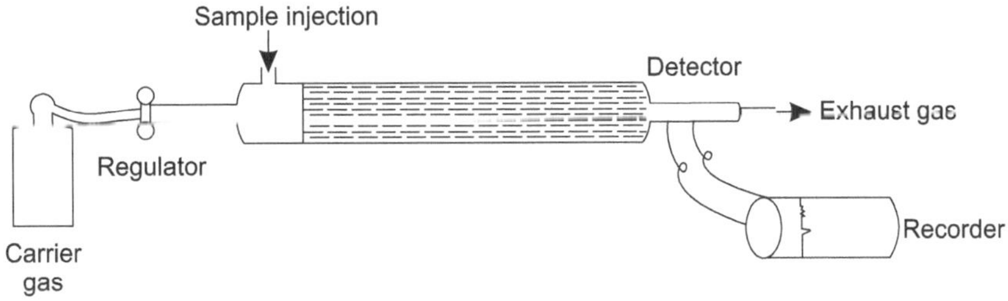

Fig. 4.20 Flow diagram of a gas chromatograph

The flow rate of the carrier gas (60 – 80 mL/min) is controlled by an regulator. The above procedure is gas-solid chromatography (GSC), the fixed phase is a solid material like granular SiO_2, Al_2O_3 or C. A common type of gas chromatography is gas-liquid chromatography (GLC). In this the fixed phase is a non-volatile liquid like silicone oil, polyethylene glycol etc, which is held as a thin layer on a solid support (diatomaceous earths or crushed fire brick, 60 – 80 mesh or 100 – 200 mesh size. The packed column is usually 4 mm (internal diameter) tube of stainless steel or copper bent in coiled form. The total length of the tube is 120 cm to 150 cm. These days, capillary column (0.025 mm inside diameter and 50 meters long) are used. These can resolve efficiently up to several hundred components in a single sample. The capillary column may be of metal, glass or organic polymer, in which walls act as support for the stationary liquid phase.

The accuracy and selectively of the GC depends on the type of detector used. The temperature of the thermostatic Jacket surrounding the column and the detector is maintained between 50°C and 500°C or varied for temperature programming, depending on the need. Different types of detectors are used. These include thermal conductivity detectors, electron capture detector, flame thermocouple detector, prizoleitric absorption detector, photoioniation detector etc.

The flame ionisation detector (FID) being very sensitive in the best for detection of organic compounds, which are pyrolysed in hydrogen-oxygen flame producing ions. These ions are helpful to carry current through the flame. The ions get collected at a charged electrode and the resultant current is measured by a suitable electrometer amplifier.

GC Analysis

Gas-liquid chromatography (GLC) is one of the most important and commonly used technique for the separation of mixture of compounds. The instrument that is used is called a gas chromatograph. The sample of the mixture (liquid, gas or solution in a suitable solvent) is introduced into the instrument at the head of the column using a micro needle. The inert carrier gas (like nitrogen, helium etc) is also introduced in the sample injection chamber, which is connected to a coiled column filled with a suitable absorbent. The flow of the carrier gas and the mixture of compounds passes through the column. The rate of movement of the various components depends on the tendency to dissolve in the stationary liquid phase. Compounds having least solubility in the stationary phase move faster through the column and the compounds having more solubility pass through the column in order of their solubility (more soluble compound is the last to pass out). The circular column is maintained at a fixed temperature (using a thermostat) or alternatively heating can be programmed. Finally it is connected to the detector and finally to the recorder, which records different peaks for each compound depending on its retention time (the compounds having least solubility has least retention time and the compound having maximum solubility has maxium retention time. The various compounds are identified on be basis of their retention times. Their Concentration is determined on the basis of peak areas.

The exact identification of a compound is affected as follows. Suppose a particular peak is due to a compound *n*-octane. So this compound (*n*-octane) is introduced along with the mixture. The peak due to *n*-octane become much bigger compared to the original peak.

The compounds can also be identified by injecting the elected fractions into FT infrared instrument or mass spectral instrument. Both the FT infrared spectrometer and mass spectrometer record the spectra of each compound as it leaves the column. Thus, the identification is simplified. A diagrammatic representation of a GLC instrument is shown below (Fig. 4.20a)

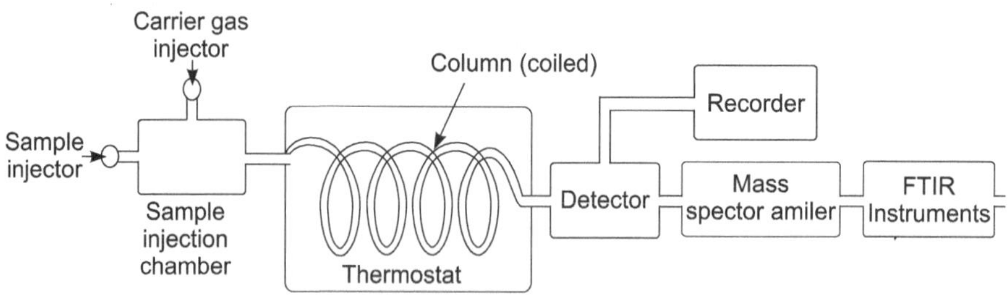

Fig. 4.20a. Diagrammatic representation of a GLC instrument

Gas Chromatography/Mass Spectrometry (GC/MS)

The coupling of gas chromatograph with mass spectrometer is a very powerful tool for organic environmental samples. Mass spectrum as obtained for all fraction of gas chromatography is helpful to identity each component unambiguously. For GC/MS only nanograms of the material is required for all information. The success of the whole operation depends in the transfer of the sample from the interface between GC/MC from high pressure operating region of GC (about 700 torr of carrier gas) to the high-vacuum conditions of the mass spectrometer (about 10–5 torr). This procedure has been successfully used for the identification of organochlorme pesticides in the environment.

High Performance (Pressure) Liquid Chromatography (HPLC)

In gas-liquid chromatography (GLC), compounds which have low vapour pressure or compounds which cannot withstand higher temperature (due to decomposition) cannot be separated for analysed in a mixture compounds. Mixtures having such compounds can, however be separated by High Performance Liquid Chromatography (HPLC). In HPLC solid adsorbents of small size particles (5-10 μm are used) and pressure up to 3000 psi is used, which helps for steady flow of the eluating solvent. The column used is a straight type and is made of stainless steel tube (about 20 cm long and a bore of 2-3 mm) to withstand high pressure. The adsorbent used depends on the nature of the solid. A diagrammatic representation of HPLC is shown in Fig. 4.21.

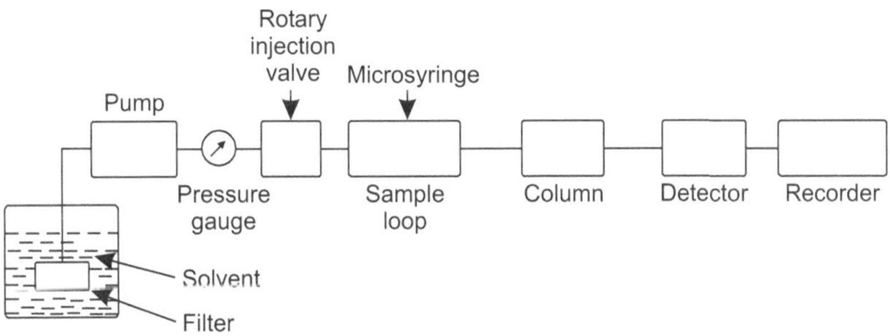

Fig. 4.21 Diagrammatic representation of HPLC

The solvent used for HPLC should be of HPLC grade. It should be extra pure and free of any suspended matter. HPLC grade solvents can be prepared by special methods of purification. A suitable pump is used to pass the solvent through the column at a pressure of 4500 psi. The sample is introduced by using a microsyringe.

The efficiency of separation of the mixture depends on the choice of the solvent or mixture of solvents. Instead of using temperature programme (in GLC) elution with solvent of gradually changing polarity is used in HPLC. The separation also depends on the temperature of the mixture.

The column can be directly connected to an IR spectrophotometer or UV spectrophotometer and spectras recorded for each component.

In case of water pollutants, the water has to be concentrated, extracted with a suitable solvent and evaporated to small volume prior to HPLC as in GLC.

Ion Chromatography

In ion chromatography, the anion are separated on a column of anion exchange resin, according to their size and charge, by elution with dil. $Na_2CO_3/NaHCO_3$

solution. The separated anions are then converted to the corresponding free acids by passing the eluate through a column of cation exchanger (H^+ form). The presence of free mineral acids, even in very low concentrations, causes a change is the conductivity of the final eluate, acid can be readily measured. Such a method, using suitable standards, provides a measure of the concentrations of individual ions.

Ion chromatography is now a valuable tool to monitor mobile and statutory sources of air pollution, water pollution in drinking water, ground and surface water, and industrial and municipal wastes. Presence of cations in acid rain and anions in industrial wastes can be measured in 10 – 15 min.

Chemiluminescence

Chemiluminescence is a standard method for monitoring NO_x, is the atmosphere. It is based on the interaction of NO with O_3 to yield electronically excited molecule, NO_x^*, with the emission of radiation in the range 600 – 3000 nm.

$$NO + O_3 \longrightarrow NO_2^* + O_2$$

The excited NO_2^* molecule thus formed loses energy and return to the ground state with the emission of energy, which is measured by a detector like a photomultiplier. The intensity of radiation is proportional is the concentration of NO_2

$$NO_2^* \longrightarrow NO_2 + h\nu$$

In the atmosphere, NO_2 is invariably associated with NO. Thus, NO_2 is first convested into NO (before analysis), which can be done by passing the air sample over a thermal converter. Analysis of this mixture gives NO_x, (a mixture of NO and NO_2). Analysis of another sample which is not passed through thermal converter gives NO. The difference of these two data gives the concentration of NO_2.

Presence of other nitrogen compounds and particulates interferes with the estimation. This interferance can be removed by using a membrane filter at the air inlet.

Chemiluminescence can also be used in the analysis of ozone by interation with NO

$$NO + O_3 \longrightarrow NO_2^* + O_2$$

X-Ray Fluorescence

This method is useful in the analysis of particulate matter and can also be used for some soil and water samples. This technique to based on the fact that when a sample is bombarded with X-ray, γ-radiation or photons, it gives rise to secondary radiations as electrons drop into vacant positions in the inner orbitals. The absorption of X-rays produces electronically excited ions that return to their ground state by transitions involving electrons from higher energy levels. The emitted X-rays are characteristic of the particular aton and yield a qualitative analysis of the elements, while the intensity of the radiation provides a quantitative analysis with reference to a standard sample. A X-ray fluorescence spectrophotometer is shown in Fig. 4.22.

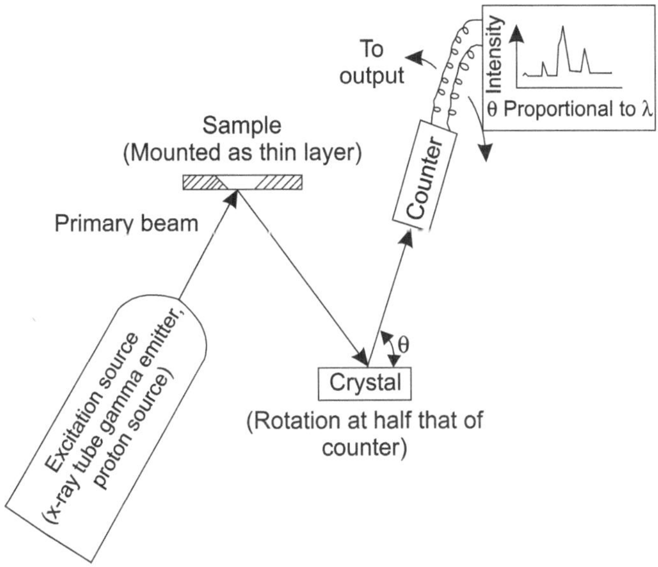

Fig. 4.22 X-ray fluorescence spectrophotometer

4.12 SUMMARY

Table 4.5 gives a summary of the various instrumental methods which are used for various pollutants.

Table 4.5 Summary of the Gaseous Air Pollutants and Particulate Matter Analysed by Different Instrumental Techniques

Gaseous air pollutant	*Instrument technique*
NO_2	Chemiluminesence, IR spectrophotometery, visible and UV spectrometry
SO_2	Spectrophotometry, conductivity
CO	Non-dispersive IR spectroscopy, gas chromatography
Hydrocarbons	IR spectrometry, GS-MS chromatography, Ionisation analysis
Aromatic hydrocarbons	GC-MS gas chromatography.
Metals	Emission spectroscopy, atomic absorption spectroscopy.
Ozone and oxidants	Visible and UV spectrometry.
NH_3	Spectrophotometry, potentiometry
Volatile pesticides	GC-MS gas chromatography
H_2S	Spectrophotometry, UV molecular absorption
Silicates	Chromatography
Sulphates	Electron spectroscopy

EXERCISE

1. What are particles, ions and radicals. How are they formed in the atmosphere?
2. Discuss the chemical and photochemical reactions of oxygen, nitrogen, nitrous oxide, carbondioxide, ozone and sulphur dioxide in the atmosphere.
3. What is Smog? How it is formed? Discuss its effects.
4. What are the major air pollutants?
5. Carbon monoxide is poisonous. Explain?
6. Carbon dioxide is not considered to be a pollutant in reasonable amount but is excess amount it is a pollutant. Explain?
7. Discuss the sources of air pollutants.
8. Write notes an

   (a) Acid rain (b) Green house effect (c) Global warming
9. Discuss the role of chlorofluorocarbons in the atmosphere.
10. Describe briefly the analytical method that can be used for measuring air pollution.

Chapter 5

Industrial Pollution

5.1 INTRODUCTION

The industries, as we know, serve the backbone for the development in terms of providing employment opportunities, increased income and better living standards. Industrial revolution of the past has resulted in the production of better crops, medicines and drugs (including antibiotics). In fact the quality of life has improved due to discovery of dyes, plastics, cosmetics and other materials of our daily use. All these developments resulted in increasing the average life expediency from 47 years in 1900 to 75 years in 1990's. However, the ill effects of all the industrial developments became pronounced with the pasage of time.

Industrial wastes resulting from all manufacturing industries are a major cause of air, water and land pollution. These discharges are contaminated with materials used in plant (starting materials), manufactured end products, intermediate products and byproducts. Besides there are discharge of gaseous effluents particularly from industries using coal or oil for heating purposes.

The industrial pollution included suspended matter, chlorides, surfactants etc. the most troublesome are sulphates, nitrates, ammonium salts, salts of heavy metals and oils etc. In water bodies, these pollutants affect the dissolved oxygen (DO), biochemical oxygen demand (BOD)and chemical oxygen demand (DO). The pollution in soil affects and alters the chemical and biological properties of soil. These pollutants enter the food chain, disturb the biochemical processes and seriously affect the living organisms. The type of pollutants depend on the type of industries. Pollutants from some of the industries are discussed below.

5.2 PAPER AND PULP INDUSTRIES

The demand of paper and its products have increased considerably. New pulping and bleaching processes have been developed. However, the effluents or wastes from these industries are mainly responsible for the pollution of water bodies.

These wastes contain two types of pollutant, viz., Suspended matter and dissolved organic substances. The suspended matter consists of dispersed solids like coarse and fine particles of bark, wood bits, etc. Nearly 50 per cent of the wood used for the manufacture of paper is rejected as waste material during the pulp and paper making process. These wood particles would sink to the bottom and produce sludge which blankets fish spawning grounds and destroy certain types of aquatic life. The dissolved organic substances include lignins, tannins, carbohydrates, salts of fatty acids, methyl mercaptan, pantachlorophenol and sodium pentachlorophenate, etc. which are harmful to fish. Besides the above, the wastes contain caustic alkalies (used for pulping), filler and coating materials (e.g., clays, calcium carbonate, and titanium dioxide), inks, sizing and hydrolysis products. The chemicals used in sizing, coating and bleaching process form extremely harmful constituents of wastes.

Suspended matter in the wastes can form bottom deposits after coming in contact with water. These are harmful to aquatic life and on decomposition produce odoriferous components. The water in the water bodies becomes opalescent and so the penetration of sunlight into water is reduced considerably, resulting in reduced photosynthetic process.

The dissolved organics in the discharged water are slowly acted upon by microorganisms and converted into simple harmless compounds. However, during this process, the dissolved oxygen in water is consumed and this results in the death of some aquatic organisms.

Chlorine is an excellent bleaching agent used in pulp bleaching, when discharged into aquatic systems, reacts with other industrial effluents to produce a number of chlorinated products; the most harmful of these is tetrachlorodibenzo-*p*-dioxin.

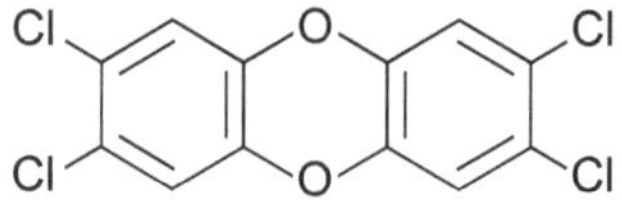

Tetrachlorodibenzo-*p*-dioxin

The environment pollution caused by paper and pulp industry is minimised by the following procedure:

(*i*) **Removal of particulate pollution.** The particles are allowed to settle down (under the influence of gravity) in settling tanks. If the suspended particles are light, coagulants such as alum or silica are added to expediate the settling process. The particulate settled material is removed from the bottom of the settling tanks. From this material, the fillers, pigments and coagulating material can be isolated and reused.

(*ii*) **Removal of dissolved organic matter.** The effluents, after the removal of suspended matter, are treated with microorganisms and oxygen in specially designed tanks. By this process, the organic effluents are biologically degraded into harmless substances.

(*iii*) **Recycling of water.** The water obtained after removal of suspended particulate matter and dissolved organic matter (as described in (*i*) and (*ii*) above) is recycled. This procedure helps to reduce the discharge of unwanted toxic chemicals in the water bodies.

(*iv*) **Use of alternative bleaching agents.** Chlorine is an effective bleaching agent but it reacts with other industrial effluents to produce toxic chemicals. In view of this, attempts have been made to replace chlorine with other bleaching agents like ozone and hydrogen peroxide. However, it adds to the cost.

(*v*) **Recycling of waste paper.** It is convenient to convert used waste paper into pulp by treatment with alkali and steam. The pulp obtained is bleached. This procedure saves a number of other chemicals which are used to convert wood into pulp.

(*vi*) **Conversion of sugars into ethyl alcohol.** In the sulphite process for the manufacture of pulp, sugars are formed by the hydrolysis of wood constituents. More than 50 per cent of these sugars are capable being of fermented to alcohol. For this, the sulphite waste is separated from the pulp and conditioned for fermentation. The solution in cooled the 30°C, its pH adjusted to 4.5 by addition of lime, subsequently urea is added as a nutrients and then fermented. The alcohol formed is distilled and rectified. In some countries like Sweden all the alcohol manufactured is based on waste sulphite liquors.

5.3 FOOD PROCESSING INDUSTRIES

The wastes from the food processing industries, such as diary forms, food preserving and meat packing unit, breweries and distilleries, etc., mostly contain organic matter which can be putrified. If such wastes are discharged into water bodies, there is depletion of oxygen. Effluents from slaughter houses may contain fats, proteins and some pathogenic bacteria. The effluents from breweries, distilleries and sugar industries contain nitrogen, sugar, proteins, fermented starch, etc. It is essential to treat such wastes before discharging into water bodies. These wastes are subjected to trickling filters to remove most of the suspended solids and also reduce the BOD. The fermented wastes after evaporation and drying can be used as animal food.

5.4 LEATHER TANNING INDUSTRY

The raw material used in leather industry (hides of cattle and skins of certain other animals) is mostly the byproduct of edible meats, which is produced in slaughter houses. The leather and tanning industry produce effluents which contain small bits of skin tissues, soluble proteins, hair, blood, dirt, salts of many kinds and lime, mixed with sulphides, chrome and vegetable tanning agents, sugars, starch, oils, fats acids, alkalies, dyes and pigments.

The tannery waste contains organic pollutants which are responsible for using the dissolved oxygen in water bodies and so are responsible for destruction of aquatic life. The pollution caused by leather tanning industry can be minimised by removal of particulate pollutants and removal of dissolved organic matter as described under paper and pulp industry.

5.5 CHEMICAL INDUSTRIES

Chemical industries include manufacturing industries like alkali (NaOH, Na_2CO_3), acid (H_2SO_4, HNO_3, HCl), fertilizer, pesticides and a number of other industries dealing with the manufacture of large number of organic (e.g. drugs, etc.) and inorganic chemicals. The wastes from most of the industries are either acidic or alkaline and must be neutralised before discharging into water bodies.

The wastes from fertilizer industries contain phosphates and nitrates, which are responsible for eutrophication of lakes and methaemoglobinaemia in infant children. The harm caused could be minimised by using lesser amounts then the recommended doses.

The wastes from pesticides industries particularly in the production of 2, 4-D release dichlorophenol into the water bodies.

5.6 METAL INDUSTRIES

The wastes from industries dealing with the manufacturing of metals contain metals like copper, lead, cadmium, mercury, chromium, zinc, etc., which are highly toxic to human beings and aquatic life. In addition to the metals, the effluents also contain acids, oils, greases and cleaning agents. The effluents from iron and steel industries contain coke, limestone, phenol, oils and suspended solids. The wastes must be treated to recover useful products, which can be reused.

5.7 CEMENT INDUSTRY

Cement industry is of great concern because of the difficulty in controlling the pollutant emission from them. The common air pollutants include particulates, coke, calcium, potassium, silica and aluminium. People in a number of countries suffer from chronic lung ailments due to particulate matter.

5.8 THERMAL POWER PLANTS

Power plants are used for generating electricity or power. In most of the conventional power plants, coal is used as the source of energy. Coal fired power plants are a source of serious environmental problems. These inject the gaseous pollutants like oxides of carbon and sulphur into the atmosphere. Besides gaseous pollutants such power plants inject particulates in the form of carbon particles and fly ash.

Fly ash is generated as waste material from thermal power stations which utilise coal as the source of energy. It creates enormous pollution problems. In India there are more than 60 thermal power plants which produce more than 30 million tons of fly ash every year. Disposal of fly ash creates air pollution, surface pollution and water pollution. A number of technologies have been developed for the manufacture of products from fly ash. These include fly ash bricks, fly ash tiles, hollow bricks, insulation bricks, etc. The use of fly ash in the manufacture of cement is also being explored.

In the thermal plants, large amount of heat goes waste. It is dissipated by condenser coils, which are cooled by water from nearby rivers and discharged back into the latter. The net result is increase in temperature of the river water by about 10°C. This is harmful for aquatic life, since it results in decrease of the dissolved oxygen. The best course is to remove thermal pollution and to put the heated water from power plants into a holding section and cool it prior to their discharge in water bodies. Alternatively, the heat generated from such units can be used for purposes like warming the building in cold seasons.

5.9 NUCLEAR POWER PLANTS

About 16 per cent of the world electrical energy is derived from nuclear energy. Nuclear energy is the energy stored in the atomic nucleus. Nuclear energy or atomic energy is used for generation of power or electricity. The plants so used are called Nuclear power plants.

Nuclear power (fission) reactors are based on the fission of U^{235} nuclei by thermal neutrons. The fission of U^{235} is represented as

$$^{235}_{92}U + ^{1}_{0}n \longrightarrow ^{142}_{56}Ba + ^{93}_{36}Kr + 3^{1}_{0}n + 200\text{ MeV}$$

The energy from the nuclear reaction is used to heat water in the reactor and steam so produced drives a steam turbine to generate electricity.

Most of the nuclear power plants use light-water nuclear reactor, which use oridinary water for transporting heat from reactor core to boilers. These reactors use $^{235}_{92}U$ as fuel. The natural abundance of this isotope is only about 0.7 per cent. The process of enrichment concentrates this isotope to about 3 per cent. As already stated, the fission of $^{235}_{92}U$ using neutrons release small fragments as ^{142}Ba and ^{93}Kr, a lot of energy and neutrons. These neutrons carry forward the chain reaction by further splitting the uranium atoms present.

Another type of reactor called **fast breeder reactor**, can increase the nuclear fuel potential by 100 times. These reactors do not require weapon-grade material and can use up to 60 per cent uranium in the ore.

The uranium reserves in India amount to 95,000 tonnes of the metal. After accounting for losses due to mining (15 per cent), milling (2 per cent) and fabrication (5 per cent), the amount available for power generation is 61,000 tonnes. But the thorium reserves are larger in quantity. The **monazite** reserves are about 8 million tonnes in which about 0.63 million tonnes of thorium is present. The amount of thorium available for nuclear power in about 2,225,000 tonnes.

In India, there are 14 operating atomic reactors accounting for power generation of 19193 million units in 2001–3. The Nuclear reactors are located in Tarapur, Rajasthan, Kakrapakkam, Narora, Kakrapara and Kaigo (two reactors in each state). Eight more reactors are under constructions at Tarapur (Maharashtra), Kaigo (Karnataka), Rawatbhata (Rajasthan) and Kudankulam (Tamilnadu).

Energy production using nuclear fuel is a clean and efficient option. Although 1kg of Uranium generates an equal amount of energy generated by 35,000 kg of coal, one has to weigh the two options (of Thermal and nuclear Power) in terms of their advantages and disadvantages (see Table 5.1)

Table 5.1 Energy from Coal and Nuclear Source

	Coal	*Nuclear Sources*
Advantages	1. Inexpensive	1. Releases CO_2
	2. Large availability	2. Small amount of solid waste is there
	3. Small capital investment	3. Large energy available
Disadvantages	1. Leads to air pollution	1. Large operational costs
	2. Releases large amount of CO_2	2. Large capital investment
	3. Surface mining disturbs	3. Problem of radioactive waste landscape disposal
	4. Transportation-infrastructure	4. Technology-availability and is costly updating
	5. Ash-disposal	5. Risk factors –Radiation hazards –Accidents

In view of the disadvantages associated with nuclear power generation, it has been a controversial issue for several years. The main problems associated with nuclear power plants are the selection of site for nuclear power plants, strict safety controls, disposal of radioactive waste and effects on human health by radiation exposure, generation of nuclear waste and its disposal.

The management and disposal of radioactive waste forms the subject matter of a subsequent section.

5.10 POLYMERS

Polymers are macromolecules which are formed by chemical combination of a large number of relatively small molecules called monomers. All plastics are polymers but all polymers are not plastics. The polymers are of two types, vig., the natural polymers like polysaccharides, proteins etc and the synthetic polymers like nylon, etc. The synthetic polymers have been successfully used in place of traditional materials like natural fibers, natural rubber, etc. They are used for clothing (synthetic fibres) and automobile tyers, tubes, etc.

5.10.1 Synthetic Elastomers

Elastomers are polymers that can be easily stretched. Rubber is the only natural elastomers known. Some degree of elasticity can be introduced by adding cross linking agents, e.g. vulcanised rubber. Since the supply of rubber from natural sources was not sufficient to meet the ever increasing demand for rubber, it was inevitable to develop synthetic elastomers, some it which are given below.

Styrene-Butadiene Rubber (Buna S)

It is obtained by the copolymerisation of styrene and butadiene in aqueous emulsion (sodium oleate as emulsifier) at 50° in the presence of potassium persulphate and dodecylmercaption.

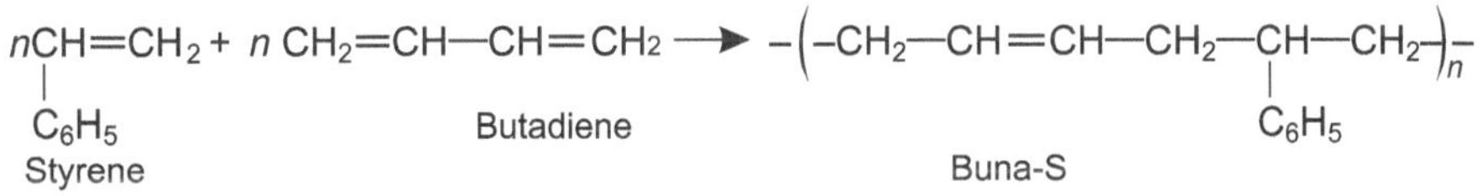

Buna-S is used for making floor tiles, coated fabrics and shoe soles, etc.

Polymerisation of styrene and butadiene at 5° in presence of cumene hydroperoxide and glucose gives another variety of Buna-S, known as SBR. It is used in the production car tyres, footwears and carpet backing materials.

Buna-N (Nitrile Rubber)

It is a copolymer of butadrene with 15 – 40 per cent acrylonitrile and is obtained by polymerisation in an aqueous emulsion using soap as an emulsifier.

Buna-N is resistant to heat, sunlight, oils and solvents and has high abrassion resistance. It is used for hose and tank lining.

Neoprene

It is obtained by polymerisation of chloroprene in aqueous emulsion at about 30° using soap as emulsifier and potassium persulphate as free radical catalyst.

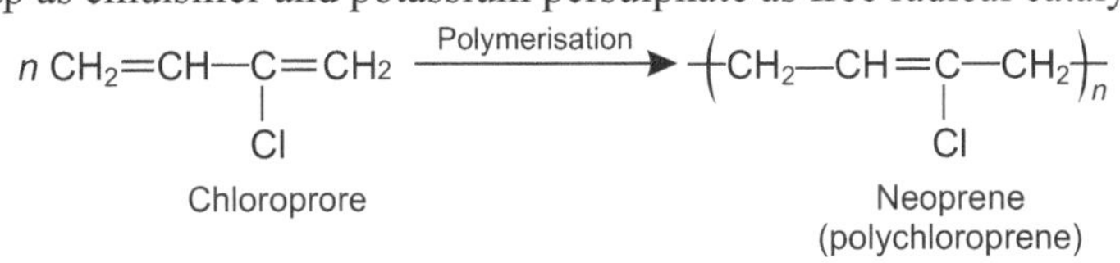

Neoprene is resistant to the action of chemicals and solvent and is used for protective clothing, linings, reaction vessels, hose covers, conveyer belts and floor tiles, etc.

5.10.2 Synthetic Fibres

Some of the fibres like rayons and cellulose acetate fibres are obtained from natural sources, e.g. cotton, silk, etc. Amongst synthetic fibres, the most important are the polyamides like nylons and the polyesters like tereline.

Nylon 66. It is obtained by the polymerisation of adipic acid and hexamethylene diamine. The polymerisation is brought about by stirring the two constituents.

$$\underset{\text{Adipic acid}}{HOOC—(CH_2)_4—COOH} + \underset{\text{Hexamethylene diamine}}{H_2N(CH_2)_6NH_2} \longrightarrow$$

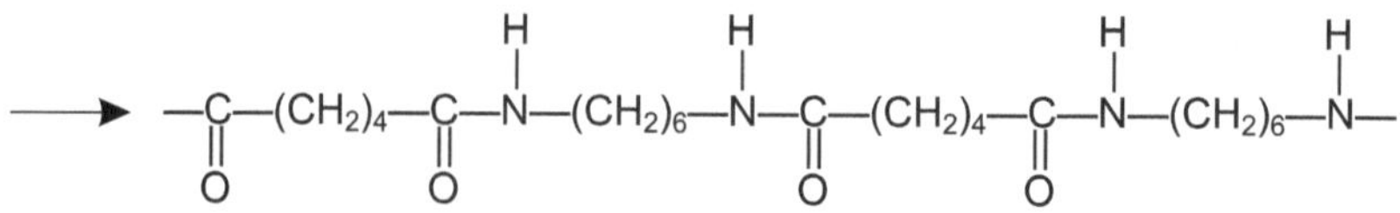

Nylon 66

Nylon 66 has high tensile strength and elasticity and low affinity for water. It is not affected by sea water and has remarkable resistance to abrasion. It is widely used for making dress material, reinforcement of rubber tyres, for brushes and brooms, manufacture of parachutes, ropes, safety belts, glider tow ropes, fishing nets, artificial fur coats and permanent pleated garments.

Nylon 610. It is obtained by the polymerisation of hexamethylene diamine and sebacic acid (C_{10} dicarboxylic acid). It is mainly used for making brush bristlers.

$$—NH—(CH_2)_6—NH—CO—(CH_2)_8—CO—$$

Nylon 610

Nylon 6. It is obtained by the polymerisation of caprolactam (which is obtained from cydohexanone).

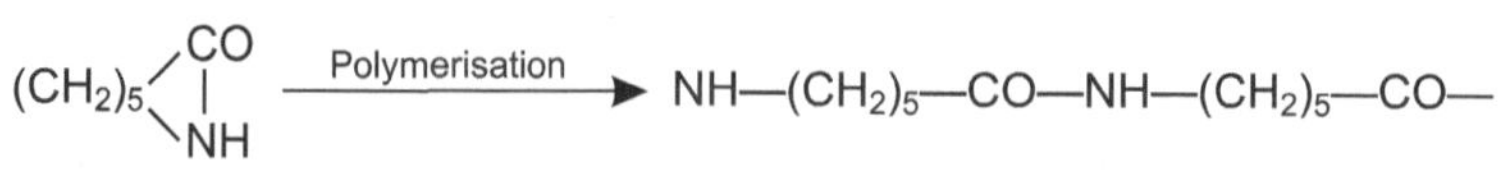

Carpolactam Nylon 6

Terelene (Dacron). It is also known as polyethelene tetraphthalate (PET) and is prepared by the polymerisation of terephthalic acid or its ester and ethylene glycol under vacuum at elevated temperatures.

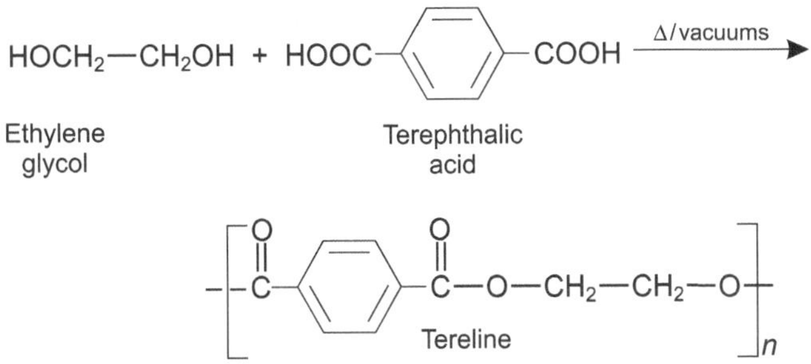

It is resistant to degradation by light and is used for making filter sheets, conveyor and safety belts, fishing nets, sails and ropes, etc.

Kodel. Its is a copolymer of terephthalic acid and cyclohexane dimethanol.

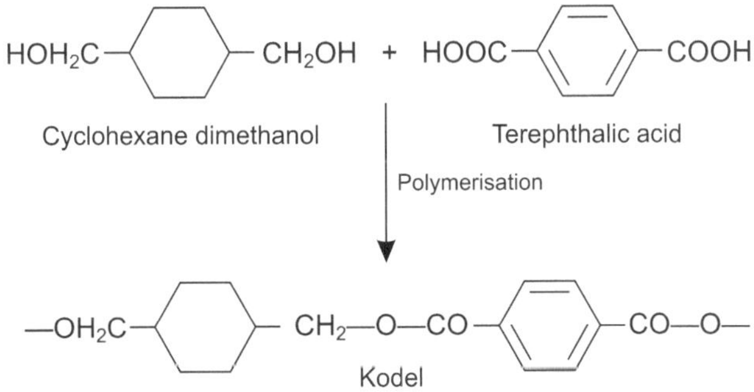

Kodel is mostly used for making dress materials.

Polymers vs Environmental Pollution

Polymers like plastics have much longer life because they are not easily biodegradable in the environment. The main reason for being not biodegradable is their hydrophobicity or water-repelling characteristics. Also, the larger molecules cannot be metabolised by microorganisms and are therefore non-biodegradable. The smaller units are easily assimilated by microbes. Thus, if by some procedure, the macromolecules could be subdivided into smaller structural entitles, for example by photodegradation, then the smaller molecules will not persist in the environment for long periods of time. Most of the synthetic fibres are polymides or polyesters, it is possible for them to undergo photodegradation (if proper environment exists) to give smaller fragments, which may be biodegraded by microbes. Also the synthetic elastomers contain oefinic bonds, it may be possible to degrade them into smaller fragments.

5.11 RADIONUCLIDE ANALYSIS

5.11.1 Introduction

Radioactive pollutants enter the water bodies (especially seas and oceans) form various sources some of which are given as follows.

(*i*) Mining and processing of ores to produce radioactive substances

(*ii*) Nuclear Power Plants

(*iii*) Nuclear Reactors

(*iv*) Nuclear Weapons

(*v*) Use of radioactive isotopes in medical, industrial and research applications

(*vi*) Testing of nuclear weapons in air, under water, underground give radioactive fall out.

5.11.2 Types of Radionuclides

Radionuclides can enter the human body through direct inhalation from air by inhalation or through the food cahin. Some radio nuclides like radon-222 and xenon-133 are absorbed on suspended particles in the air and when air is inhaled, these find their ways into the lungs, where they release the energy produced by their decay.

Through soils, radionuclides are taken up by the plants and subsequently taken up by humans when they eat the plants or consume meat of animals who have consumed such plants. Table 5.2 lists some of the radio nuclides which tend to accumulate in various human organs.

Table 5.2 Some Radionuclides and the Organs they Effect

Radionuclides	*Organ*
Sulfur-35	skin
Potassium-42	Muscles
Cobalt-60	Liver
Krypton-85	Ovaries
Strontium-90	Bones
Iodine-131	Thyroid
Polonium-210	spleen
Radon-222	Lungs
Radium-226	Bones

Nuclear power Plants generate liquid waste (low level radioactivity), liquid and gaseous wastes from fuel elements, fission products and heat. The discharge of radioactive wastes in air and water bodies cause problems in the long run. Table 5.3 lists some radioactive materials discharged from various sources.

Table 5.3 Radionuclides Discharged

Activities	*Radionuclides discharged*
Leaching of minerals	^{236}Ra, ^{40}K
Atomic reactors	^{90}Sr, ^{131}I, $^{137}C_5$, ^{140}Ba, ^{141}Cs, ^{85}Kr, ^{60}Co, ^{54}Mn, ^{55}Fe, ^{239}Pu
Neutron bombardment of atomic fuel	Radioactive Pu, Np, Am, Cm,, Bk, Cs, Zn, Ru, etc.

5.11.3 Effect of Radionuclides

The level of radionuclides found in water is measured in units of pico curies, 10^{-10} curie, (3.7×10^{-3} disintegration per second) [A curie is the quantity of radioactivity contained in 1g of radium]. Following are given some of the bad effects due to radioactive pollutants in water

- They cause serious skin cancer, carinoma, melanoma, breast cancer, leukemia, breakage of DNA and cataracts.
- They cause the so called radiation sickness, which is characterised by nausea, vomitting, diarrhoea, anorexia, epilation, lethargy and weakness.
- They cause genetic disorders, gene mutations and blood abnormalities.
- The radionuclides present is water get accumulated in soil sediments, water and aquatic system.

5.11.4 Analysis of Radionuclides

The radionuclides are released into the environment due to various activties like nuclear explosions etc. These can be analysed by counting the alpha activity (which is due to the traces of transuranic elements like U and Pu) and beta activity (which is due to radioactive alkaline earths and rare earth nuclides). Follwoing techniques are used:

- Evaporation Mcthod
- Separation Technique
- Sequential Technique

1. Evaporation Method

Water samples containing radionuclides are collected in special containers made of polythene. Water is evaporated to dryness under infrared lamp. The content is transferred to planchet and alpha activity is determined using uranium as a reference source. The beta redioactivity is determined in a Geiger-Mullar counter. By the evaporation method beta redioation with energies greater than 300 KeV and minimum concentration of 15 pCi per litre can be measured. For determination of beta activity potassium (K-40) is used as a reference source.

2. Separation Technique

This technique is used when the Evaporation method cannot be used due to low activity requiring large volume of water for analysis or when the water sample gets contaminated with excess of dissolved solids.

In the seperation technique, the dominating radioelements among alpha and beta-isotopes are separated to count their activity. Following procedure is used:

(*i*) The alkaline and rare earth fission products such as Sr-90, Ce-141, etc., are precipitated as oxalate using calcium carrier. The activity of the precipitate is termed as gross beta activity . However, radionuclides like Cs-137, I-131 and other radionuclides form complexes with oxalate radical, cannot be determined for their total activity.

(*ii*) Alpha emitters are precipitated on $BiPO_4$ in HNO_3 medium and gross alpha activity determined in the percipitate. For this, bismuth phosphate is dissolved in HCl and the gross alpha activity carried on lanthanum carries (5mL), whereby these are precipitated as lanthanum fluoride and HF. The precipitate is washed, transferred on planchet, dried and counted. By this procedure, only the gross activity can be determined. It did not indicate the specific radionuclides.

3. Sequential Technique

In this technique, the radionuclides are determined in groups as given below:

(*i*) The water sample containing Cs-137 and I-131 is filtered using Whatman filter paper No. 42. To the filtrate is added a solution of HNO_3, Cs, Ca, and KI. Iodine is reduced to iodide by the addition of hydroxylamine hydrochloride. The solution is stirred with ammonium molybdophosphate (AMP) and AgCl is added. The precipitate of Cs-AMP and AgI-AgCl are collected on a planchet, dried and beta activity counted.

(*ii*) The filtrate is treated with Na_2HPO_4 and NH_4OH. The precipitated calcium phosphate is filtered, dissolved in HNO_3 and the volume made to 5 mL in a standard flask. One mL of this solution is evaporated and gross alpha and beta activity counted.

(*iii*) From the remaining 4 mL portion, bismuth phosphate ($BiPO_4$) is precipitated to estimate Ra and U. For this, $BiPO_4$ is dissolved in HCl and precipitated as LaF_3 by adding La-carrier. The precipitate is evaporated and alpha activity counted.

(*iv*) From the above filtrate, uranium is extracted with ether and estimated fluorimetrically.

(*v*) The aqueous fraction (from (iv) above) is acidified with HCl followed by addition of Pb and Ba carriers. EDTA helps to separate $PbSO_4$ from $BaSO_4$. Barium sulphate carries radium from which alpha activity is counted.

5.12 DISPOSAL OF WASTES AND THEIR TREATMENT

Waste is produced in all countries. The amount of wastes to be managed is enormous. Broadly speaking wastes are of two kinds. The one is that generated from municipal (including domestic) wastes and the other that is generated from various industrial proceses. Industrial processes generate a variety of waste products, which are generally discharged into water bodies or even on land. The nature of industrial wastes depends on the industrial process in which they originate. In general, the industrial wastes contain organic or mineral matter along with other corrosive, poisonous and other harmful substances, which are injurious to health. The organic matter of the wastes are destroyed slowly by the bacteria.

A common practice that is followed in some industries is storing or lagooning wastes. Air is passed through the wastes. By this process the organic wastes are degraded. Also if the waste contains acid as well as basic wastes, it reduces the cost of neutralisation. This process also decreases the suspended matter and also results in the reduction of biochemical oxygen demand BOD.

It is advantageous to treat the wastes at the source with a view to recover valuable byproducts. Fermentated wastes after evaporation and drying are used as animal food. Metals like chromium from electroplating industries can be recovered by the use of ion exchangers.

5.12.1 Types of Industrial Wastes

Industrial wastes are classified into a number of types. These include **process wastes** and **chemical wastes**. In the process wastes, every industry produces its own waste. These arise from different operations such as washing of the raw materials purification of the intermediates and final products. The process wastes are of two types:

1. Inorganic process wastes. Such wastes emerge from inorganic chemical manufacturing industries, metallurgical industries, petroleum industries. The inorganic wastes do not pose many biological problems provided the toxic metals are not present.

2. Organic process wastes. Such wastes arise from food processing industries, dairies, breweries, distilleries, paper mills, textile mills and organic chemicals manufacturing industries.

Chemical wastes arise from industries manufacturing acids, bases, detergents, explosives, dyes, agro chemicals. Some chemical waste require biological oxidation treatment methods, such as trickling filters, activated sludge or lagooning.

5.12.2 Treatment of Wastes or Effluents with Inorganic Impurities

Such wastes are subjected to chemical treatment. The pH of the effluents is maintained between 6.0 and 9.0. The treatment depends on whether the effluents are acidic or alkaline. If acidic, the wastes are neutralised by lime (CaO). The alkaline water (as from tanneries, textiles and chemical industries) are normally treated with H_2SO_4. After neutralisation, the effluents may be discharged into water bodies.

In case the effluents contain heavy metals, there may be precipitated by using suitable precipitating agents. Certain waste water may contain cyanide—such wastes are subjected to chlorination under alkaline conditions (chemical oxidation) or with ozone.

5.12.3 Treatment of Wastes or Effluents with Organic Impurities

Such wastes are subjected to conventional biological treatments. Before these treatments, the wastes are neutralised, waste matter is removed by aeration. Suspended organic pollutants and colloidal impurities are coagulated by adding $FeSO_4$ or alum or some other flocculating agent. After the primary treatment the effluents may be subjected to biological treatment such as **activated sludge** and **trickling filter treatments**. In case the number of microorganisms in the effluents are less, requisite amount of microorganisms have to be added.

Special types of problems arise when we deal with **Hazardous wastes**. There have been several practices for managing such wastes. Some of such practices are:

- Production of less waste.
- Conversion of hazardous substances into less hazardous ones.
- Safe and proper disposal of wastes.

1. Reduction in the amounts of waste

It is safest and the least expensive option. The amount of hazardous waste generated by manufacturing and other processes can be reduced by changes in the chemical processes involved, equipment used, raw materials used, etc.

Recycling and *reuse* also eliminates hazardous waste. For example, certain solvents and acids can be recovered and recycled. This practice yields economic benefits both in terms of reducing cost of starting materials and costs involved in the handling of hazardous waste.

Also certain industries can mutually benefit by taking wastes from one industry and using it by another industry as raw materials.

The waste produced in practically all industries can be prevented or at least minimised to a bare minimum by following the Principles of green chemistry.

2. Conversion to less Hazardous Substances

The toxicity or hazardous characteristics can be reduced by *physical, chemical* or *biological methods*.

Physical methods such as absorption on charcoal or resin filters for absorbing toxins can be used. Distillation is also useful at times. Precipitation and immobilisation are also helpful in making wastes suitable for long-term storage.

Chemical methods such as neutralisation, oxidation, reduction, ion exchange, etc., detoxify the waste.

Biological methods involve the use of bacteria, soil microorganisms and other species in reducing the hazardous characteristics of the waste.

Biodegradation can be accomplished by aerobic and anaerobic methods.

Incineration is a physical method of treatment. It is very important and has many advantages. It involves heating of the waste to 1200°C for quite sometime to complete destruction. The volume of the waste is reduced by at least 90 per cent and the resulting waste is safer to store in a landfill or other storage site. Incineration removes volatile, combustible and mobile organic matter from the wastes by converting them to CO_2, water and other gases. It is a quick, easy and permanent solution for many hazardous wastes and can be used for even mixture of wastes.

3. Disposal Methods

The final and ultimate step in the management of hazardous wastes is its permanent storage. But after storing it, we should be able to regularly monitor its safe placement by drawing the samples. The two important methods of disposal are *secure landfills* and *deep—well disposal.*

***(i)* Secure Landfills:** Land filling has been a very common and popular option for disposal of hazardous waste. But, the landfills should be secure. The modern *secure landfills* are specially designed for safe disposal of the waste. Such a landfill is shown in Fig. 5.1

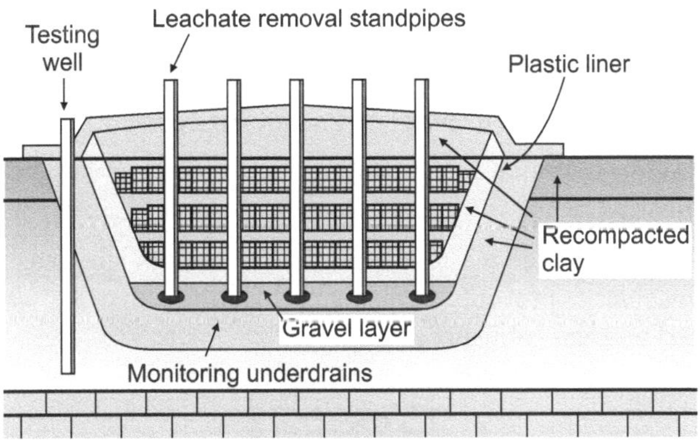

Fig. 5.1: A schematic diagram of a secure landfill

The bottom of the 'secure landfill' is lined with compacted clay followed by a layer of gravel. A thick polythene liner is then placed over the ground layer. The hazardous wasts are packed in drums and put into the landfill. The filled landfill is then finally covered with clay, plastic and soil.

***(ii)* Deep Well Disposal.** It involves injection of waste into deep wells. The waste in injected into a permeable rock layer several thousand meters below the surface, in geologic basins topped by relatively impervious, fracture resistant rock. The method requires careful monitoring after disposal.

Although oceans dumping was used for disposal of hazardous wastes but due ot the consequences of marine pollution, this practice has been stopped.

Biomedical waste. It comprises of used needles, syringes, blood bags, urine bags, glucose bottles, surgical gloves, bandages, chemicals, body parts and radioactive substances. Such a kind of waste is generated in hospital, nursing homes, pathological labs and research institutions.

The biomedical wastes possess special problems of spreading disease such as AIDS, hepatitis, fevers, septicaemia and other infections. It has a potential danger for workers, public and the environment.

The biomedical waste (Handling and Management) rules, 1998 prescribe the segregation of waste at source, colour coding of waste bags, incineration of infectious waste, careful disposal of waste and shedding of plastics to prevent their reuse. According to the handling and management rules, the treatment of biomedical waste involves processes such as incineration, autoclaving, microclaving, hydroclaving and shredding.

- **Incineration** is a controlled combustion process where waste is completely oxidised and harmful microorganisms present in it are destroyed or denatured by high temperature.
- **Autoclaving** is a low-heat thermal process where steam is brought into direct contact with waste in a controlled manner for sufficient duration to disinfect the waste.
- In **microclaving**, microbial inactivation is done by thermal effects of electromagnetic radiation having frequency between 300 and 3,00,000 MHz.
- **Hydroclaving** is similar to autoclaving except that the waste is subjected to indirect heating by application of steam in the outer jacket.
- **Shredding** cuts the waste into smaller pieces to make the waste unrecognisable. It prevents its reuse and signifies that the waste has been disinfected and can be disposed off.

EXERCISE

1. What pollutants are caused by the following industries?

   (*a*) Paper and pulp industries (*b*) Chemical industries

   (*c*) Thermal power plants (*d*) Nuclear power plants

   (*e*) Polymers.

2. How can environmental pollution caused by the following industrial units be removed.

   (*i*) Paper and pulp industries

   (*ii*) Chemical industries

   (*iii*) Polymer industries

   (*iv*) Thermal power plants

   (*v*) Nuclear power plants

3. What are radionuclides. How do they effect the environment?
4. Write a note on disposal of industrial wastes and their management.

Chapter 6

Noise Pollution

6.1 INTRODUCTION

Modern life style has contributed to a new form of pollution called **noise pollution**. It is caused by over crowded cities and towns, the transport, devices for recreation and entertainment. Other factors that contribute to noise pollution include horns by motor vehicles, using loudspeakers and film songs at various functions. Though all these factors are inevitable part of modern life style, they are responsible for disturbing human peace. Thus noise has become an environmental pollutant and effects environmental health. Technological developments like development of industry, machinery, etc., create noise.

Noise is unpleasant, unwanted or disagreeable sound that make people uncomfortable. Noise pollution is unwanted sound released in the environment without regard to the adverse effects it may cause.

6.2 MEASUREMENT OF NOISE

The unit of measurement of noise is decibel (dB), which is one tenth of bel, which is named after Alexander Graham Bell. A bel is the 'logarithm of a ratio' of any two sound intensities. Thus, 1 decibel = 10 bel = $10 \log_{10}\left(\frac{I}{I_0}\right)$, where I is the intensity

of sound which we want to measure and I_0 is the intensity of the reference sound

One dB is equal to the faintest sound, a human ear can hear.

Table 6.1 gives some common sounds and their characteristics

Table 6.1 Some common sound and their characteristics.

Source	*dB*	*Intensity (Loudness)*	*Effect*
Threshold of hearing	0	Hardly audible	
Breathing	10	Just audible	
Whispering	20	Very quiet	
Library	30	Quiet	
Bedroom	40	Quiet	
Business office	50		
Conversation (average)	60	Moderate	Loud
Vacuum Cleaner	70	Loud	Annoying
Alarm clock	80	Very Loud	May damage hearing
Food processor, Diesel Truck	90		Hearing damage
Jet air craft at 300 m	100	Uncomfortable	
Steel mill	110	Loud	
Automobile pressure horn	120		
Thunder	130		Pain
Jet take off at 100 m	140	painful	
Jet take off	150	,, ,,	Effects eardrum
Rocket engine	180	,, ,,	Effects eardrum

6.3 SOURCES OF NOISE

Noise can originate from natural or man made sources. The natural sources include thunder, sea-shores etc. Exposure to natural noise sources is very less. However, noise originated from man made sources are important and need attention.

(*i*) Industries

A number of industries like iron and steel, automobile, power plants, petroleum, fertilizers, etc., involve operation like drilling, mining, blasting, etc. The workers in these industries are exposed to noise levels of 80-90 dB.

(*ii*) Domestic Appliances

Domestic appliances like coolers, air-conditioners, washing machine, vacuum cleaner, food processors, etc., generate noise. This noise has become a routine for us.

(*iii*) Transport

Noise pollution is caused by both air and surface transport. Taking off of jet planes generates a lot of noise. People living in the vicinity of airports are worst effected.

In big cities surface transport like cars, buses, trucks, two wheelers, etc., contributes to noise pollution. Even railway tracks contribute to noise pollution. The average noise levels are approximately 90 dB.

(*iv*) Construction Activities

Construction bridges, metro rail services, flyovers, subways, etc. use large equipments like cranes, mixers, etc., at the construction site. People living in the vicinity of construction sites are the worst effected.

(*v*) Religious Activities and Festivals

During religious activities in places of worship, loudspeaker are invariably used, which are responsible for noise pollution. Crackers are used during Deepawali and Dussehra festivals; these crackers not only pollute the air but also generate loud noise.

6.4 EFFECTS OF NOISE POLLUTION

According to WHO, noise can affect human health in a number of ways. These include physical, physiological and psychological effects.

(*i*) Physical Effects

These are general fatigue, high blood pressure leading to cardiovascular changes and digestive problems. Prolonged exposure to noise levels more than 80 dB may cause deafness.

(*ii*) Physiological Effects

Noise pollution affects the biological functioning of the body. The affects include anxiety, insomnia, hypertension, nausea, giddiness, etc. The respiratory system can be affected by noise levels in the range 120–150 dB.

Due to high noise level, the secretion of hormones of the pituitary gland, viz. adrenocorticotropic hormone (ACTH), which in turn stimulates the adrenal gland and there is further secretion of a number of other hormones. The net result is that the sensitivity of the body to adrenalin increase in the blood sugar level, Suppression of the immune system and malfunctioning of the liver.

Noise levels of about 70 dB may cause constriction of the blood vessels in skin. Frequent exposure to noise levels of more than 85 dB can lead to higher pulse rate and higher cholesterol levels. Chronic noise may also lead to abortion and congenital defects.

(*iii*) Psychological Effects

Noise can affect sleep, which in turn results in overall mental and physical health of a person. Even noise of low levels can cause annoyance to some people and can lead to bad temper, quarrel or enmity. Even animals in the zoo are very sensitive to noise, their behaviour can be affected.

6.5 CONTROL OF NOISE POLLUTION

Noise pollution can be minimised by using some control measures. Vehicular noise can be reduced by discarding the old vehicles and replacing them with latest design vehicles. In a similar way noise pollution in industries can be reduced by changes in design of machinery and equipment. Same rule applies to appliances used in constriction activities.

It feasible, the residential areas should be at a distance from railroads, airports, highways and industrial establishments.

Sound absorbing material like glass wool, felts, porous blocks and perforated files and carpets, etc., can also reduce noise.

Workers in industries may use ear-plugs or ear-muffs. Trees absorb and dissipate energy and act as buffer zones. Planting of trees along highways, streets and near industrial establishments in helpful to reduce noise pollution.

Noise pollution can best be controlled through laws. This can help to prevent and control pollution in industrial, commercial and residential areas. An example is having silence zones near hospitals, schools, etc.

Manufacture and bursting of crackers can only be controlled by strict laws.

EXERCISE

1. Why is noise considered to be a pollutant.
2. What is the difference between noise and sound.
3. What are the effects of noise pollution on health.
4. How can noise pollution controlled.

Chapter 7

Environmental Toxicology

7.1 INTRODUCTION

Toxicology is the science of the study of poisons. Originally developed by Paracelsus, toxicology is concerned with the investigation of the deleterious effects of all foriegn substances on living organisms. **Environmental toxicology** is the branch of science which concerns with the nature, characteristics, effects and detection of toxic substances in the environment and any speiecs which is exposed to the environment. The environment refers to surroundings and is defined as the sum total of all living (abiotic) and non-living (biotic) components.

The study of environmental toxicology involves the occurrence, availability and form of toxic chemicals in the environment; exposure of such chemicals and the harm they cause; and mechanism of action on the exposed species including humans. The toxic chemicals are discharged primarily via industries into all the segments of the environment, viz., atmosphere, hydrosphere and lithosphere. Some of these chemicals enter into the human body via food chains from the environment. Once they enter our biological system, various biochemical process are affected. This may lead to adverse effects including death.

7.2 CHEMICALS IN THE ENVIRONMENT

There are a large number of chemicals in the environment. Some of these are toxic and the rest non-toxic. Many metals, which are listed as environmental hazards, are essential dietary trace elements required for normal growth and development of living beings. These elements include Al, Sb, As, Be, Bi, Cd, Co, Cu, In, Pb, Mo, Ag, Te, Sn, Ti, W, U and Zn. In such cases, it is the amount or the concentration of the respective element which makes it toxic. Even the well-known toxic elements like As, Pb and Cd are required in trace quantities for the growth of animals. A typical example is that of aluminium, which is biologically inert, but it causes brain damage, bone disease and anemia in patients subjected to haemodialysis using water containing 100–1000 parts per billion of aluminium. In fact 100–1000 parts per billion is the normal concentration of aluminium in drinking water where alum is used for water treatment.

The fact is that everyone is exposed everyday to toxic chemicals at home, at work, at school and in the outdoor environment. The workplace is an important source of exposure. Thus, roofers, construction workers and automobile engineers are exposed to particulate matter besides other pollutants. Farm workers are exposed to pesticides. Any one who cleans anything may be exposed to chlorinated solvents. Office going people come in contact with photocopiers, toners, correction fluids and rubber, cement. Hospital workers are surrounded by disinfectants. Everyone is exposed at home to cleaners, detergents, carbon monoxide and paints. Animals, plants and microorganisms encounter many of these substances as environmental pollutants.

It is in fact, because of the presence of toxic chemicals in the environment that environmental toxicology is important.

7.2.1 Toxic Chemicals in Air

The usual major pollutants (like oxides of carbon, nitrogen, sulphur), particulate pollutants and metallic pollutants which pose serious health hazards, it is necessary to control these pollutants. Besides the above, the US Environmental Protection Agency along with consumer products safety commission has identified 24 extremely hazardous substances in the atmosphere. These are: acrylonitrile, arsenic, asbestos, berellium, cadmium, chlorinated solvents, chlorofluorocarbons, chromates, coke oven emissions, diethylstilbestrol, dibromochloropropane, ethylene bromide, ethylene oxide, lead, mercury, nitrosoamines, ozone, polybrominated biphenyls, polychlorinated biphenyls, radiation, sulphur dioxide, vinyl chloride and toxic waste disposal emissions and leachates.

7.2.2 Toxic Elements in Water

A list of toxic trace elements found in natural and waste water is given in Table 2.20. Some of these are essential at low levels (necessary as nutrients for animal and plant life), but are toxic at higher levels. Besides the toxic elements, the water bodies contain a large number of pesticides, which are primarily agricultural run offs.

7.2.3 Toxic Waste in Soils

Soil has been used as a dumping ground for most of the waste products including toxic wastes. The main sources of these wastes are

- Industrial wastes
- Urban and Domestic wastes
- Radioactive wastes

- Agrochemicals like fertilizers and Pesticides
- Plastics and Polymers
- Chemical and metallic Pollutants
- Biological agents.

7.3 BIODEGRADABILITY

It is a common knowledge that a number of pollutants in the environment (e.g., some pesticides, plastic and polymers etc.) are non-biodegradable. This means that such pollutants remain in the environment for a very long time and their hazardous effects are felt for long periods of time. On the other hand, biopolymers like starch, cellulose, proteins and nuclic acid etc. are biodegradable.

The term **biodegradation** is commonly used for **biotransformation** processes. All kinds of living organisms, from bacteria to humans can convert organic chemicals and many inorganic chemicals to other simpler substance, which in most of the cases have no harmful effects in the environment. For example, the vapor phase oxidation of volatilised chemicals cleans the atmosphere. Also photo reduction destroys water pollutants. In fact, biotransformations are the primary self cleaning mechanisms for several pollutants. Biotransformations help living organisms to make environmental toxic substances into less toxic or non toxic substances. Biotransformations are also helpful for recycling the nutrients.

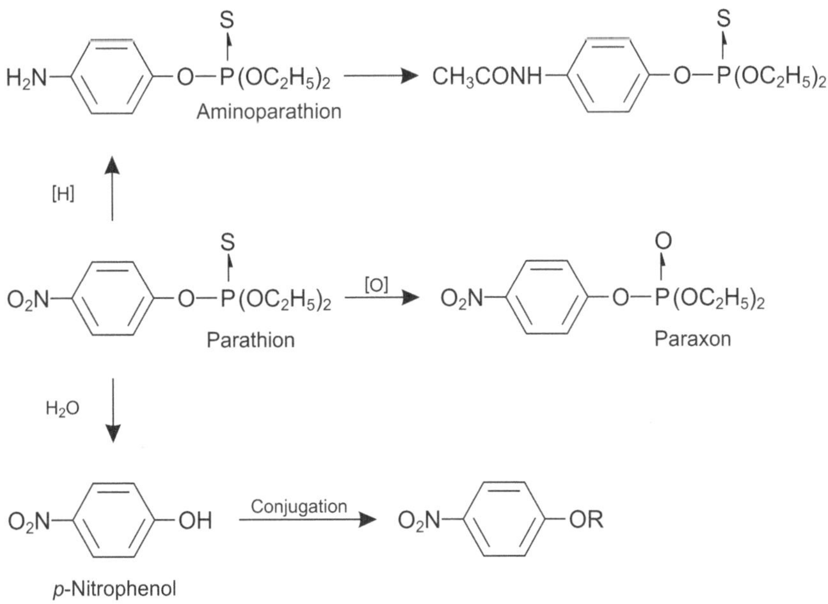

Scheme 7.1

A typical example of a biotransformation or biodegradation is the urine analysis of a person who has been exposed to traces of the insecticide parathion. It is found that the urine sample contains the metabolites amino-parathion (obtained by reduction of nitro to amino group) and paraoxon (obtained by S-oxidation) and *p*-nitrophenol (obtained by ester hydrolysis). The amino-parathion can be converted into *n*-acetyl derivative. All the products are formed enzymatically (see scheme 7.1).

In the above biotransformation the reduction of NO_2 is achieved by FAD or NADPH and biochemical oxidation of parathion to Paraxon utilise the familiar oxidants, perferryl (FeO^{3+}) or percupryl (CuO_2^{2+}) ions and hydrolysis is catalysed by a protein rather than a mineral base or acid. Synthesis (conjugation) involves high energy reagents just as in the laboratory. Amino acetylating is achieved by acetyl coenzyme A, the biochemical equivalent of the chemist's acetyl chloride.

Table 7.1 gives the major biotransformations alongwith principal reactions and reagents that we normally come across for biodegradations.

Table 7.1 Some Major Biotransformation Reactions.

Type	***Reaction***	***Typical enzyme***	***Co factor***	***Reactant***
Oxidation	C—C → C—C (epoxide, bridging O)	Aldrin epoxidase	NADPH	FeO_3^+
Reduction	$-NO_2 \rightarrow -NH_2$	Nitro reductase	NADPH	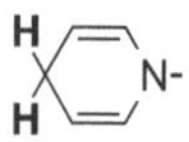
Dehydrochlorination	H–C—C–Cl → C=C	DDT dehydro-chlorinase	GSH[a]	**HS**-CH_2CH
Hydrolysis	-COOR → -COOH	Lipase	Serine	**HO**-CH_2CH
Acylation	$-NH_2 \rightarrow$ -NHCOR	Arylamine acyl-transferase	Acyl CoA	-CH_2**S-COR**
Methylation	$-OH \rightarrow -OCH_3$	O-Methyltrans-ferase	SAM[b]	CH_3 over $CH_3S\text{–}CH_2^-$
Glycosidation	R-OH → R-OX	UDP transglucur-onylase	UDPGA	X-O-PPU
Sulfation	$R\text{-}OH \rightarrow R\text{-}OSO_3^-$	Sulfotransferase	PAPS[c]	$^-$O–S(O)(O)–OPA

[a] GSH glutathione, P phosphate, X sugar, U = uridine

[b] SAM S adenosylmethonine

[c] PAPS[1] 3′-phosphoadenosane A – 5′-phosphosulfate, A = adenosyl

7.4 PRINCIPLES OF DECOMPOSITION (BIODEGRADATION)

These include biooxidation, bioreduction and biohydrolysis

7.5 MICROBIAL TRANSFORMATIONS

Most of the biodegrations are brought out by microorganisms. They include bacteria, actinomycetes, protozoans, fungi, yeasts and algae.

7.5.1 Biooxidations

Amongst microbial transformations, bioxidations are the most significant. Biooxidation is a form of burning, but one that takes place at ambient temperature and is enzyme catalysed. A typical example is the biooxidation of oil spills. When oil spills, the lower hydrocarbons volatilise into the atmosphere and microbial oxidation removes most of the rest. The higher alkanes present in oils are degraded by enzymes. The microbial oxidation of a long chain alkane is represented as shown below (Scheme 7.2).

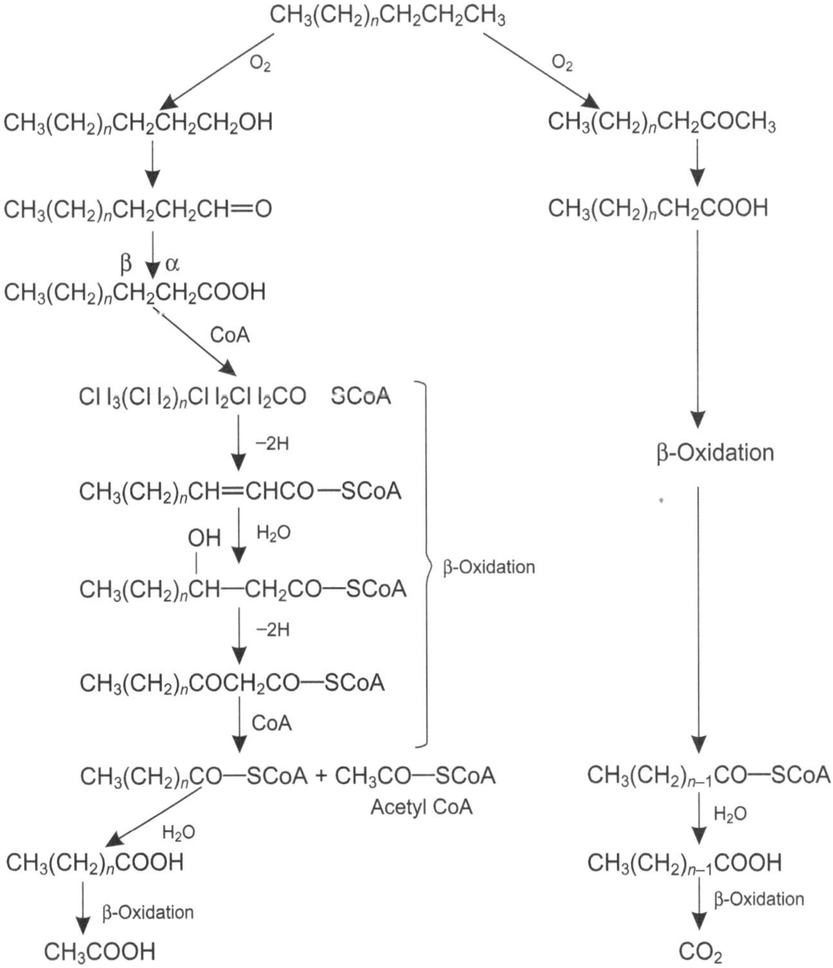

Scheme 7.2 Microbial oxidation of alkanes, CoA = coenzyme A

P-oxidation does not work in branced chain alkanes. This is the reason why in the earlier, branced alkyl benzenesulphonates which were used widely as detergents (due to their low cost) were not biodegradable. Conversion to n-alkyl detergents solved the problem of nonbiodegradability. Benzene rings are oxidised by bacteria by atmospheric oxygen across a double bond leading to the ultimate formation of catechol. Also, other organic compounds, such as benzoic acid are biotransformed to catechol by the same route (scheme 7.3)

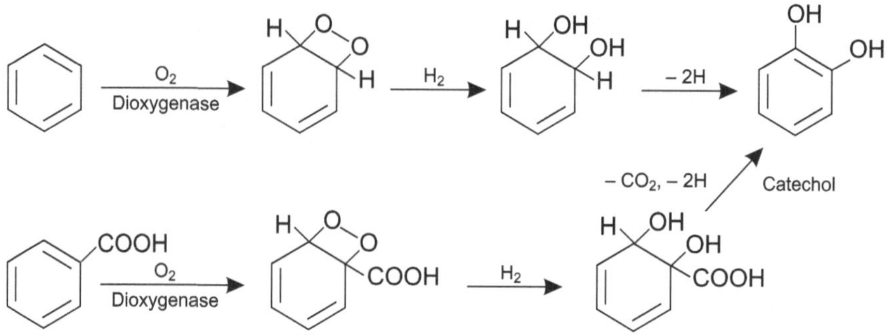

Scheme 7.3

The catechol ring is subsequently cleaved by oxidation along either of the two routes (scheme 7.4)

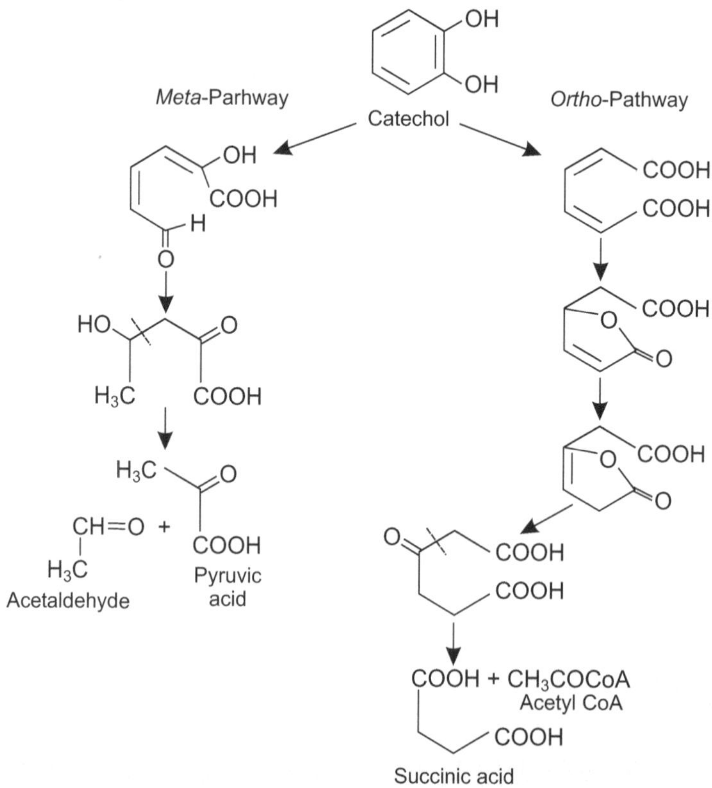

Scheme 7.4 Microbial cleavage of catechol

Even polycyclic aromatic hydrocarbons like naphthalene, anthracene and phenanthrene undergo microbial oxidations to catechol (scheme 7.5) which in turn is biodegraded.

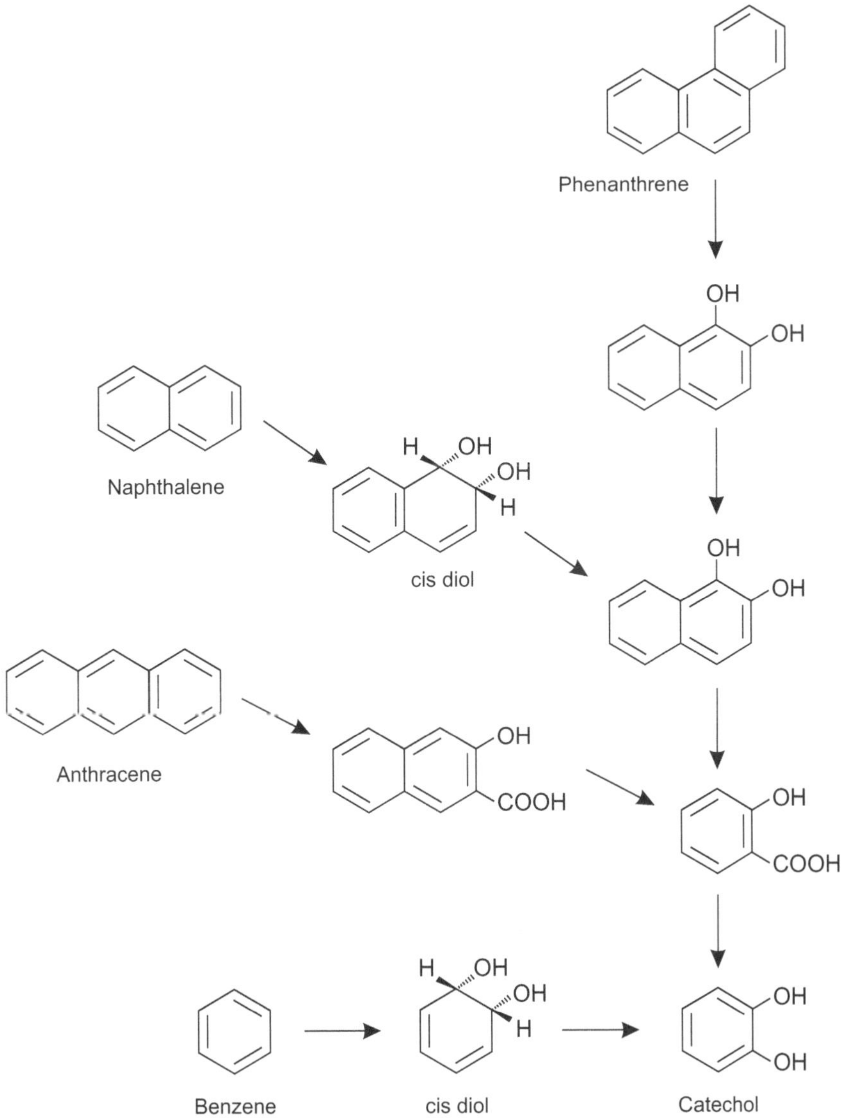

Scheme 7.5 Microbial oxidation of aromatic compounds

The bacterial oxidation shown in scheme 7.5 are catalysed by dioxygenases. Some other hydroxylations, epoxidants etc. can also be achieved by mono-oxygenase system as in **cytochrome 450**, an iron-containing enzyme (its designation arises from the uv absorption maximum at 450 nm in its reduced and co-complexed form). Cytochrome 450 also oxidises naphthalene to naphthols.

All biological oxidations involve oxidases. The same products can be obtained by **dehydrogenation**, even in the absence of molecular oxygen. Thus, alcohol can be dehydrogenated by *alcohol dehydrogenase* and further oxidised by *aldehyde dehydrogenase* to carboxylic acid (scheme 7.6).

$$\underset{\text{Ethyl alcohol}}{CH_3CH_2{-}OH} \xrightarrow[\text{Dehydrogenase}]{-2H} CH_3{-}CH{=}O \xrightarrow[\text{Dehydrogenase}]{-H_2,\ +OH} CH_3{-}COOH$$

Scheme 7.6

It is by the above process (scheme 7.6) that wine is converted to vinigar. The accumulation of a acetaldehyde is in fact responsible for the pleasure of wine to hangover. The conversion of ethyl alcohol to acetaldehyde can also be accomplished by P450.

7.5.2 Bioreductions

Many bioreductions such as conversion of aromatic nitro compounds to primary amines are also mediated by cytochrome P450. The reduction take place in discrete steps (scheme 7.7), the intermediates can be isolated.

$$\underset{\text{Nitro}}{R{-}NO_2} \xrightarrow{2[H]} \underset{\text{Nitroso}}{R{-}N{=}O} \xrightarrow{2[H]} \underset{\text{Hydroxylamine}}{R{-}NHOH} \xrightarrow{2[H]} \underset{\text{Amine}}{R{-}NH_2}$$

Scheme 7.7

Although bacteria do not utilize cytochrome P 450 for such reductions, they have similar iron-sulphur enzymes and are powerful reducers.

Use of bioreductions as a degradative process is in the field of pesticides. Some of the pesticides like DDT are relatively stable in aerobic soils but are rapidly reduced under anaerobic conditions.

7.5.3 Biohydrolysis

A large number of microorganisms are effective in the hydrolysis of esters, amides, epoxides and acetals. A typical example of an amidase is penicillins (β – lactamase), the hydrolytic enzyme that deacetylate the antibiotic penicillin.

7.6 IMPACT OF TOXIC CHEMICALS ON ENZYMES

We have known that different elements have different role to play in biological system. (see section 1.9 and Table 1.5). The biological role of certain metals has also been discussed earlier (see section 1.9).

In general, toxic chemicals attack the active sites of enzymes, inhibiting their function. Heavy metal ions, particularly Hg^{2+}, Pb^{2+} and Cd^{2+} act as enzyme inhibitors. They have affinity towards sulphur containing ligands, e.g. SCH_3 and SH in methionine and cysteine amino acids, which are part of the enzyme structure (scheme 7.8).

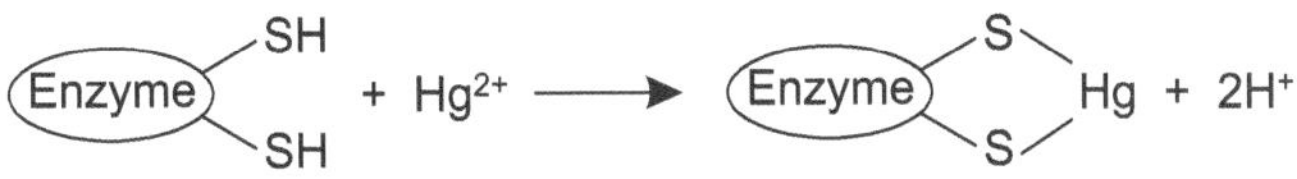

Scheme 7.8

The action of metalloenzymes is inhibited when one metal ion of a metalloenzyme is replaced by another metal ion of similar size and charge. Thus, Zn^{2+} in some metalloenzymes is substituted by Cd^{2+} leading to cadmium toxicity. In fact, cadmium inhibits the activity of adenosine triphosphate, alcohol dehydrogcnasc, amylasc, carbonic anhydrase, peptidase activity in carboxypeptidase, and glutamic-oxaloacetic transminase. Also lead (Pb^{2+}) inhibits acetylcholanesterase, alkaline phosphate, adenosine triphosphate, carbonic anhydrase, cytochrome oxidase and same enzymes in the synthesis of heme.

Biochemical effects of same toxic metals like arsenic, cadmium, lead and mercury is given below.

7.6.1 Biochemical Effects of Arsenic

The insecticides, fungicides and herbiacides contain arsenic. Among the compounds of arsenic, those of As (III) are most toxic.

As (III) exerts its toxic action by attacking SH groups of an enzyme (as in case of Hg^{2+} scheme 7.8) thereby inhibiting enzyme action (scheme 7.9).

Scheme 7.9

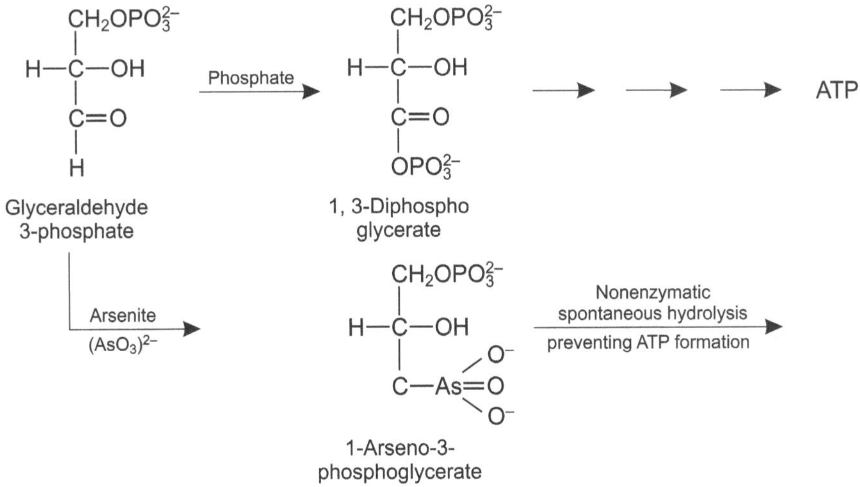

Scheme 7.10

As (III) deactivates the enzyme pyruvate dehydrogenase by complexation with —SH groups and in this way the generation of ATP is prevented. Being similar to P, arsenic interferes with the generation of ATP (adenosine triphosphate). Normally ATP is obtained from glecereldehyde -3 -phosphate as shown in scheme 7.10. However, in presence of arsenite, 1-arseno-3-phosphoglycerate, is formed, which prevents ATP formation.

In biological systems, As (III) compounds at high concentration coagulates proteins by attacking the sulphur bonds maintaining the remaining part of the structure.

The general antidote for arsenic poisoning are chemicals having —SH groups, which can bind to As (III). One such examples 2, 3-dimercaptopropanol

$$\begin{array}{c} HSCH_2\text{—}\underset{\displaystyle |\atop \displaystyle SH}{CH}\text{—}CH_2OH \end{array}$$

7.6.2 Biochemical Effects of Cadmium

Cadmium is found to occur in nature along with zinc ores. When the plants acquire zinc, they also take up cadmium. When animals eat the plants or when the human eat the animals which have consumed plants containing cadmium, the cadmium gets accumulated in human bodies. The first case of cadmium poisoning was reported in Japan in the form of 'itai itai' or 'ouch ouch' disease. A large number of people suffered from the disease in which their bones became fragile. Other effects of cadmium in the human system are Renal Dysfunction, anemia, hypertension, bone marrow disorder, and cancer.

The cadmium ingested in our body is trapped in the kidneys and gets eliminated. However, a small fraction is bound by the body proteins, metallothionein, present in the kidney. With age, the accumulation of cadmium in the body increases. When excessive amounts of Cd^{2+} are present in the body, it replaces Zn^{2+} at key enzyme sites.

7.6.3 Biochemical Effect of Lead

Compounds containing As (III) coagulate proteins. The effect on human depends on the arsenic dose. For details see chapter 2 (page 51 and 67).

7.6.4 Biochemical Effects of Mercury

Mercury, one of the most well known toxic metal is responsible for causing genetic defects in new born babies whose mothers consumed fish contaminated with mercury. For more details see chapter 2 (pages 68-70).

7.6.5 Carcinogens

The term carcinogen is used for chemicals which can cause cancer in animals and humans. The carcinogens affect the DNA, which is responsible for giving necessary directions for the normal cell growth. Table 7.2 lists a number of inorganic and organic carcinogens which are known to cause cancer, as per recommendations of US occupational safety and health administration.

Table 7.2 Some Carcinogens

Chemical	*Use/Source*	*Type of Cancer*
Inorganic		
Arsenic	Insecticide, herbicide sprays	Lung, Skin
Asbestos	Brakelining, fire proofing	Mesothelioma
Cadmium	Electroplating, Vinyl Stabilizers, pigments	Prostate
Chromium	Chrome plating	Lung, nasal
Radon	Mines, underground structures	Lung
Organics		
Benzene	Solvent	Marrow
Benzidine	Manufacture of dyes, polymers	Bladder
Benzo (α) pyrene	Cigarette and coal smoke	Lung
Bis (chloromethyl) ether	Plasticizer	Lung
Creatinine	Cigarette smoking	Lung
2-Naphthylamine	Optical bleach	Bladder
Vinyl chloride	manufacture of PVC	Liver

7.7 SOME ENVIRONMENTAL EPISODES

7.7.1 Bhopal Gas Tragedy

On the night of 2nd December, 1984, the World's worst ever industrial disaster took place in Bhopal, India. It is also known as the Hiroshima of the chemical industry because of the death toll and the suffering it caused, 27 tones of methyl isocyanate and other deadly gases were released from the Union Carbide's pesticide manufacturing factory in Bhopal. The methyl isocyanate was used in the manufacture of the insecticide *carbaryl* which was marketed under the commercial name Sevin. The sequence of the reactions involved is as follows.

$$\underset{\text{Methyl amine}}{CH_3NH_2} + \underset{\text{Phosgene}}{COCl_2} \longrightarrow \underset{\text{Methyl isocyanate}}{CH_3{-}N{=}C{=}O} + 2HCl$$

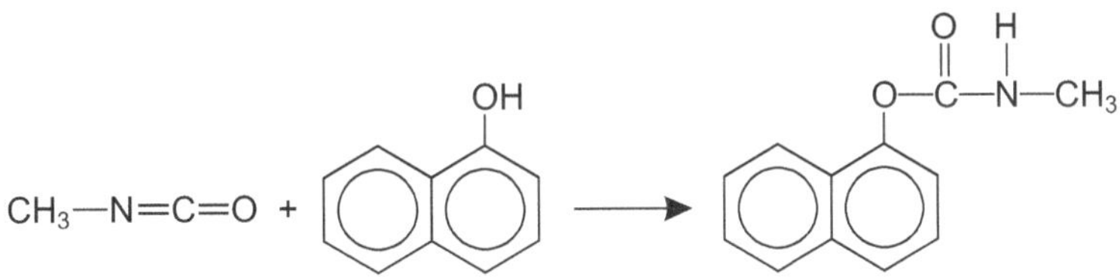

Carbaryl

Methyl isocyanate, which was stored in the tanks, leaked and the cloud of the gas gradually spread over a large area of the city killing about 4000 people and affecting thousands of others. Approximately half a million people were exposed to the leaked gases and at least 22,000 have died so far because of the effects of the exposure. It is estimated that about 120,000 people are suffering from serious long-term health effects. The number of deaths as reported at that time is much below the actual number because the deaths continued to occur due to the long-term effects of the exposure to the gas. In fact, one person died per hour because of these long-term effects of exposure.

The leakage of the gas caused irritation in the eyes, nose and mouth of the people. The people felt the burning sensation and some of them felt as if some one was frying chillies. They started coughing and froth streaking up with blood from their mouths. They started running here and there. Some had convulsions and died while the others lost control over their senses and urine and faeces fell from their legs while they were running. Even, the people ran over the dead bodies of the others.

It seems that water had entered into the largest tank containing methyl isocyanate (MIC). Unfortunately, none of the six safety systems were operational and plant siren had been turned off. The vent gas scrubber was also turned off otherwise the leaking gas could have been detoxified. The pipe connecting to the Flare tower had also been removed for maintenance. The refrigeration system for cooling liquid MIC was shut down in June 1984 to save money. Out of the three storage tanks, tank E 619 was empty while the tank E 610 contained 40 tonnes and E 611 contained 15 tonnes of the gas. On that disastrous night, water entered into the tank E 610 and an uncontrollable exothermic reaction began. The tank was blown out of its concrete sarcophagus and a deadly cloud of MIC, hydrogen cyanide, monomethyl amine and other chemicals spread out over the city of Bhopal.

Methyl isocyanate (MIC) is always associated with unreacted phosgene, $COCl_2$ to the extent of about 2 per cent. The threshold limiting value (TLV) for MIC is 0.02 ppm and for $COCl_2$ is 0.1 ppm. Exposure to MIC results in tightness in the chest and breathing troubles due to irritation of the respiratory tract. Since MIC is invariably accompanied by $COCl_2$, the combined effect becomes fatal within a span of 24hrs. Phosgene is known to be deadly poisonous gas and was used as a poison gas during World War I. The immediate symptoms of MIC—$COCl_2$ poisoning are bronchospasms, coughing, constriction and chest pain. Majority of the victims (about 80 per cent) die in the first 24 hrs. and those surving may die from broncho-pneumonia or are disabled for the rest of their lives.

7.7.2 Chernobyl Disaster

The worst nuclear accident in history occurred on April 26, 1986 at the 1000 megawatt Reactor IV at the multiunit Chernobyl Power station near Kiev on the Black sea. A violent explosion had followed a series of experiments. The explosion brusted through a 4000 tonne steel concrete cover. The reactor core temperature reached more than 2000°C. The graphite in the core caught fire and melted part of the fuel. The pressure of hydrogen built up led to the explosion, which removed the top of the building above the reactor. The fire produced a cloud of radioactive particles which went about 5000 ft. in the atmosphere.

The radioactive cloud spread north-west wards and affected a large area of Ukraine, Byelorussia, Russia and Northern Europe. When high levels of radioactivity were recorded by workers of the nuclear power plants in Sweden, the accident became known to the public.

Among the radioactive elements released were cesium-137 and iodine 131. Iodine-131 is a substance that can enter the human diet by falling on grasses that milk producing cows can eat and then concentrate in the human thyroid gland. Estimates of total radiation from fall out have been varied, one source compared it to the total of radiation from all atmosphere weapon tests. According to another source, the radiation released to the atmosphere compared that from about 300 atomic bombs.

Once it was realized that the fire could not be controlled by conventional means, helicopters were used for dumping sand, boron, dolomite and clay on to the open reactor. These neutrons absorbing materials slowed the fissioning of Uranium-235 and decreased the production of heat and radioactivity. Subsequently several foot layer of concrete was used to entomb the reactor and isolate it from the ground water and the nearby river. The studies on long-term effects on exposure to radiation are still going on.

It was reported that 237 people suffered from acute radiation sickness and 31 people died. Approximately 115,000 people were evacuated in the 30 km. Zone. About 24,000 people received an average radiation of 0.43 Sv. However, a total of about 3 billion people in the Northern Hemisphere received varying amounts of radiation. It is estimated that the radiation exposure would cause about 16,000 more deaths in 50 years time due to long-term effect of radiation. It has also been observed that a number of people suffered from disease like leukemia, thyroid, cancer etc. due to radiation exposure. The vegetation area within 7 km. of the power plant was severely damaged. The contamination of soil and water made the area inhabitable.

Several questions have been raised about the safety and proper handling of the nuclear power plants. Even with the best of the solutions, the risk of accidents would always be there.

7.7.3 Three Mile Island Disaster

The reactor incident at the three Mile Island (TMI) Nuclear Power Station in middle town, Pennsylvania occurred on 28th March 1979. The incident occurred due to cutting off of cooling water supplies due to failure of the two pumps in the secondary cooling water loop of reactor 2. The emergency pumps which are used under such circumstances were not functional due to repairs etc. This caused the overheating of the reactor which came close to meltdown. However, the situation was controlled and 100,000 people had to be evaluated from the surrounding areas. The accident released radioisotopes and radiation of 1mSv into the environment. The radiation was intense near the site. Even on third day after the incident, 12mSv/hour was recorded near the site. Although no one died as a direct result of the TMI accident, it could have been escalated into as event for worse as happened in Chernobyl in 1986. This accident created an awareness about the importance of tackling emergency situations like this and better management of nuclear power plants.

7.7.4 Minamata Disaster

Minamata disease was caused by mercury poisoning. For details, see chapter 2 (pages 68 to 71).

7.8 BETTER INDUSTRIAL PROCESS

7.8.1 Introduction

The advancement of technology has led to the new manufacturing process industries for the production of newer drugs, medicines, dyes, plastics, cosmetics and numerous other material. Due to all these developments, the quality of life on earth has become much better. For all these developments, chemistry or chemical process are responsible. However, virtually all industrial manufacturing process are responsible for environmental pollution including pollution of air, water and land. This is attributed mainly due to the effects of byproducts of chemical industry and industrial effluents, which are being discharged in air, water bodies and the land. The solutions to all these problems is to have better industrial processes, so that there is least or no environmental pollution. This is possible by invoking the concepts of green chemistry. This forms the subject matter following discussions.

7.8.2 Green Chemistry

Green chemistry is defined as environmentally benign chemical synthesis. The synthetic procedures are designed in such a way that there is virtually no pollution to the environment. We know that as on today, maximum pollution to the environment is caused by numerous chemical industries. The cost involved in the disposal of the waste or byproducts is also enormous. In view of this, it is most advantageous to design synthesis for manufacturing processes in such a way that the waste products are minimum and that these should have no effect on the environment. Also, the disposal of the waste products, if any, should be convenient. With this view in mind, the synthetic procedures should be designed in such a way that the starting materials are used up to the maximum extent to give the final product. The reactions should not generate any toxic byproducts.

7.8.3 Designing a Green Synthesis

Any target molecule can be synthesized by a number of alternative routes. The method of choice should not use toxic starting materials and as far as possible, should eliminate byproducts and wastes. Following considerations should be kept in mind.

Choice of Starting Materials

The starting materials should be selected with extreme care. The synthetic procedure will depend on this. Keep in mind the hazards that may be faced by the workers (chemists carrying out the reactions) and also the transporters who handle the starting materials.

Most of the starting materials make use of petrochemicals (made from petroleum), which are non-renewable. Also, petroleum refining requires lot of energy. It is important to reduce the use of petrochemicals by using alternative starting materials, which may be of biological/agricultural origin. For example, some of the agricultural products such as corn, potato, Soya and molasses can be transformed using a variety of processes into products like textiles, nylon etc. Some examples of materials of biological origin (obtained from biomass) are butadiene, pentane, benzene, toluene, phenolics, aldehydes, acetic acid, acrylic acid, sorbitol, glucose, furan, tetrahydrofuran etc.

Choice of Reagents

The choice of the reagent should be made on the basis of its availability, efficiency and its effect on environment. Different reagents can the used for a target molecule but the nature of the byproducts, per centage yield etc may vary.

Choice of Catalysts

It is well known that using catalysts, certain reactions proceed much faster at a lower temperature giving good yields. Catalysts like heavy metal catalysts should be avoided since they cause environmental problems besides being toxic in nature. The feasibility of a reaction to proceed under photochemical conditions should be explored. If possible, biocatalysts (enzymes) should be preferred for various steps.

Choice of Solvents

Some of the solvents which are used are extremely harmful. For example, benzene—a commonly used solvent is known to cause or promote cancer in humans and other animals. Certain other aromatic hydrocarbons like toluene can damage brain and may have adverse effects on speech, vision or cause liver and kidney problems.

Halogenated solvents commonly used, e.g. methylene chloride, chloroform, polychloroethylene and carbon tetrachloride have been identified as suspected human carcinogens.

The well-known CFCs (chlorofluorocarbons) were used up to 20th Century as cleaning solvents, blowing agents, for molded plastics and for refrigeration. Though these have very low toxic effects to humans, they are non-inflammable but are responsible for depletion of the ozone layer.

An excellent and versatile solvent is carbon dioxide in liquid state or supercritical CO_2 fluid. The later is most commonly used as a solvent. A gas normally is converted to a liquid state by increasing the pressure exerted on it. However, if a substance is placed at a temperature above its critical temperature, Tc (31°C for CO_2) and critical pressure, Pc (72.8 atm. for CO_2), a supercritical fluid is obtained. The Tc of a substance is the temperature above which a distinct liquid phase of the substance cannot exist, irrespective of the pressure applied. Pc is the pressure at which a substance no longer can exist in gaseous state. In supercritical liquid, due to high pressure, the individual molecules are pressed so close together that they are almost in liquid state. Supercritical liquids have density close to that of liquid state and viscosity close to that of gaseous state.

It is best to carry out the reactions in aqueous phase. It is now possible to carry out a large number of reactions, (which were earlier carried out in anhydrous solvents) in aqueous phase. The use of water as a solvent has distinct advantages.

Another way to carry out the reaction is without the use of solvent. In some cases, the starting materials and the reagents serve as solvents. Alternatively, the reactions can be performed in molten state to ensure proper mixing. One well-known example is that of Claisen rearrangement, which is carried out by heating without any solvents. There is still another way in which reactions can be carried out on solid support. All such procedures do not need any solvent.

7.9. BASIC PRINCIPLES OF GREEN CHEMISTRY

Green chemistry as has already been stated (see.7.8.1) is defined as the environmentally benign chemical synthesis. Any synthesis performed on small scale or industrial scale should create none or minimum byproducts which pollute the environment. According to Paul T. Anastas and John C. Warner (Green chemistry, Theory and Practice, Oxford University Press, New York, 1988), the following basic principles of green chemistry have been formulated.

- Prevention of waste/byproducts
- Maximum incorporation of the reactants (starting materials and reagents) into the final product
- Prevention or minimization of hazardous products
- Energy requirement for any synthesis should be minimum
- Selecting appropriate solvent
- Selecting appropriate starting materials
- Use of protecting groups should be avoided when ever possible
- Use of catalyst should be preferred whenever possible
- Products obtained should be biodegradable

- The manufacturing process should be so designed so as to eliminate the possibility of accidents during operations
- Strengthening of analytical techniques to control hazardous compounds

7.9.1 Prevention of Waste/Byproducts

The synthesis should be carried out in such as way so that formation of waste (byproducts) is minimum. It is very important, since in most cases, the cost involved in the treatment and disposal of waste adds to the overall cost of production. Even the unreacted starting materials form part of waste. The waste or byproducts, if discharged in the atmosphere, water bodies or land, not only cause pollution but also require expenditure for cleaning up.

7.9.2 Maximum Incorporation of the Reactants into the Final Product

It is believed that if the yield of a reaction is more than 90%, the reaction is considered to be good. The per centage yield is calculated by the expression.

$$\%\ \text{yield} = \frac{\text{Actual yield of the product} \times 100}{\text{Theoretical yield of the product}}$$

A perfectly efficient synthesis in which one mole of a starting material gives one mole of the final product (making the yield of 100%) is considered to be extremely efficient by the above calculations. However, the above procedure may generate significant amount of waste or byproducts, which is not visible in the above calculation. Such a synthesis is by no means a Green synthesis. Typical reactions like Wittig reaction and the Grignard reaction illustrate above contention. Both these reactions may proceed with 100% yield but do not take into account the considerable amounts of byproducts obtained.

A synthesis or a reaction is considered to be *green* if there is maximum incorporation of the starting materials and the reagents in the final product. One should take into account the per centage atom utilization, which is calculated or determined by the following equation.

$$\%\ \text{Atom utilization} = \frac{\text{Mol. wt. of desired product} \times 100}{\text{Mol. wt. of (desired product + waste products)}}$$

The concept of atom economy developed by B.M. Trost (Science 1991, 254, 1471 – 1477) is a consideration of how much of the reactants end up in the final product. The same concept was determined by R.A. Sheldon (Chem. Ind. (London). 1992, 903 – 906) and is given as

$$\%\ \text{Atom economy} = \frac{\text{Formula weight of atoms utilized} \times 100}{\text{Formula weight of reactants used in the reaction}}$$

Let us consider some common reactions, viz. rearrangement, addition, substitution and elimination reactions to find out, which ones are more atom economical.

Rearrangement Reactions

In these reactions, there is rearrangement of atoms. The most well known example is Claisen rearrangement.

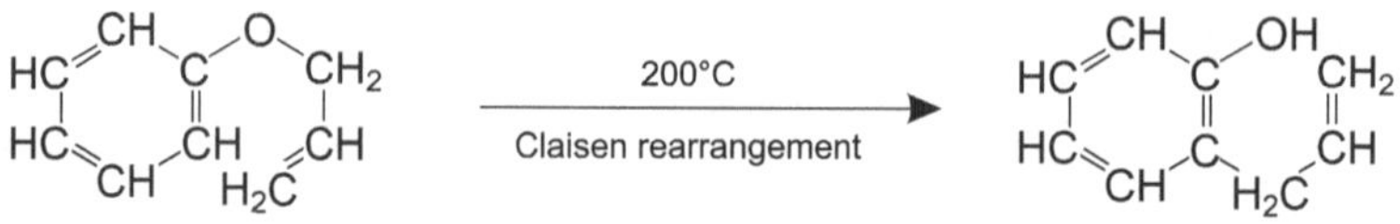

Allyl phenyl ether 2-Allyl phenol

This is a 100% atom economical reaction, since all the reactants are incorporated into the product.

Addition Reactions

The addition of H_2 to an olefin is an addition reaction

$$H_3C{-}CH{=}CH_2 + H_2 \xrightarrow{Ni} CH_2CH_2CH_3$$

Propene Propane

In this example also, all elements of reactants (propene and hydrogen) are incorporated in the final product (propane). This reaction is a 100% atom economical reaction.

Similarly, addition of bromine to olefins and cycloaddition reactions of olefins are 100% atom economical reactions.

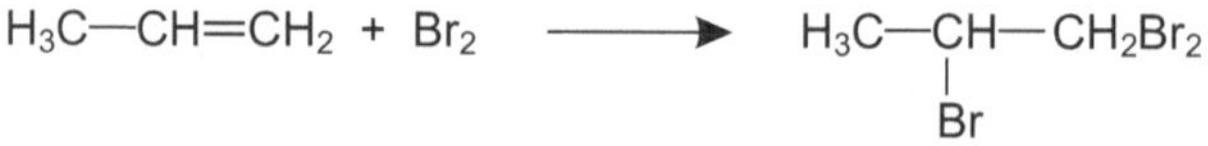

Propene 1, 2-Dibromopropane

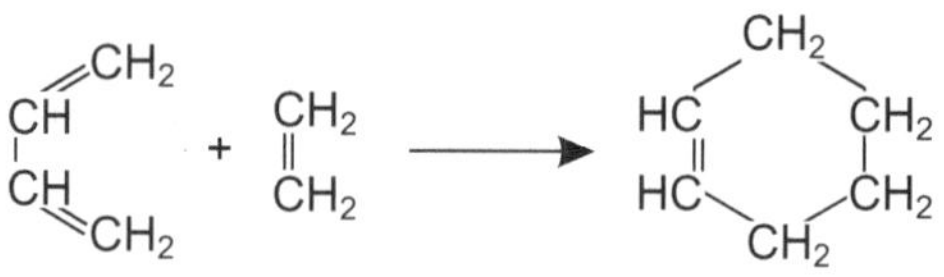

Butadiene Ethene Cyclohexene

Substitution Reactions

In these reactions, one atom or group of atoms is replaced by another atom or group of atoms. The atom or group that is replaced is not utilized in the formation of the final product. Thus substitution reactions are not 100% atom economical; these are less atom economical that rearrangement or addition reactions. Consider the reaction of ethyl propionate and methylamine.

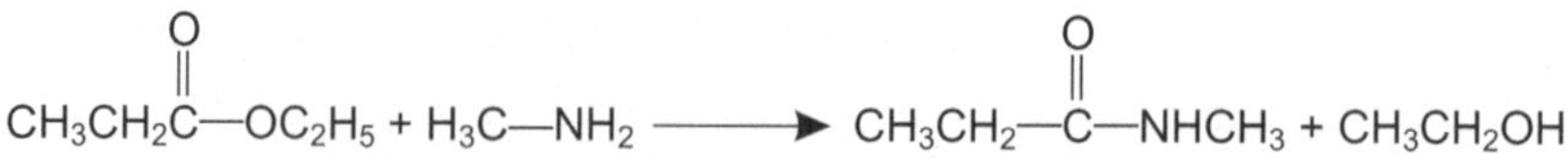

Ethyl propionate Methyl amine N-Methylpropionamide Ethyl alcohol

In the above reaction, the leaving group (—OC_2H_5) of the ester and one hydrogen atom of methylamine are not incorporated in the final product. The total of atomic weights of the atoms in the reactants that are utilized is 87.106g mol^{-1} while the total molecular weight including the reagent used is 133. 189g mol^{-1} (see Table 7.3). Thus, a molecular weight of 46.069g/mol^{-1} remains unutilized in the reaction.

Table 7.3

Reactants		*Utilized*		*Unutilized*	
Formula	*FW*	*Formula*	*FW*	*Formula*	*FW*
$C_5H_{10}O_2$	102.132	C_3H_5O	57.057	C_2H_5O	45.061
CH_5N	31.057	CH_4N	30.049	H	1.008
$C_6H_{15}NO_2$	133.189	C_4H_9NO	87.106	C_2H_5OH	46.069

$$\% \text{ Atom economy} = \frac{87.106 \times 100}{133.189} = 65.40\%$$

Elimination Reactions

In an elimination reaction, two atoms or group of atoms are lost from the reactant. Consider the following Hofmann elimination reaction.

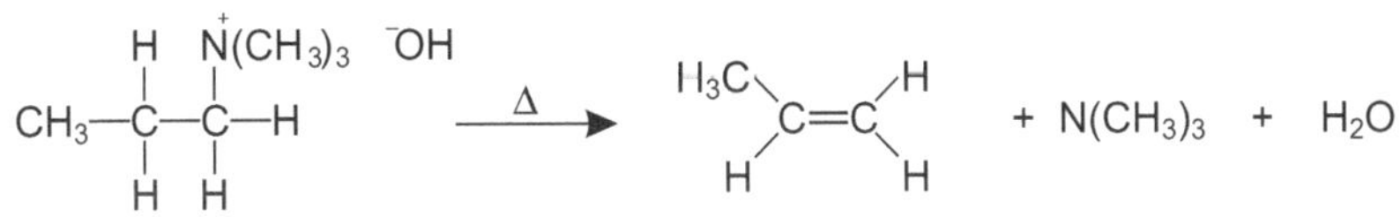

Trimethyl propyl ammonium hydroxide — Propene — Trimethyl amine — Water

The above elimination reaction is not very atom-economical. The percentage atom economy is 35.30 per cent. In fact, this is the least atom economical of all the above reactions.

Let us consider another elimination reaction involving dehydrohalogenation of 2-bromo-2-methylpropane with sodium ethoxide to give 2-methylpropene.

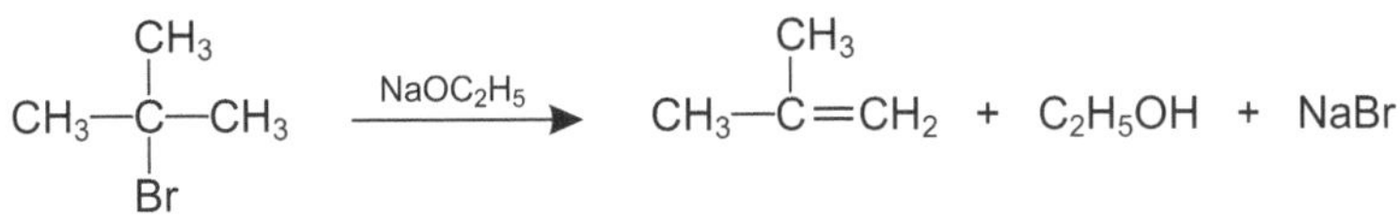

2-Bromo-2-methylpropane — 2-methylpropene

The above elimination reaction is also not very atom-economical. The percentage atom economy is 27 per cent which is even less that Hofmann elimination reaction.

7.9.3 Prevention or Minimisation of Hazardous Products

An important basic principle of Green chemistry is to prevent or at least minimise the formation of hazardous products, with may be toxic and environmentally harmful. These hazardous products may affect the workers. The minimum that should be done is to protect the workers by protective clothing, respirators etc. This, of course, will add to the cost of production. Also, the controls may fail putting the workers at greater risks. Green chemistry offers a scientific option to deal with such situations.

7.9.4 Designing Safer Chemicals

It is extremely important that the chemicals synthesized or developed (e.g., dyes, adhesives, paints, pharmaceuticals etc.) should be safe to use. An example of an unsafe drug is thalidomide (introduced in 1961) for reducing the effects of nausea and vomiting during pregnancy (morning sickness.) The off springs born to women taking this drug suffered from the birth defects including missing or deformed limbs. The use of thalidomide was banned and strict laws were formulated for testing new drugs.

With the advancement of better technology, it has now become possible to design and produce safer chemicals. Chemists can manipulate the molecular structure to achieve this objective.

7.9.5 Energy Requirements for Synthesis

For any synthetic procedure, the energy requirements should be kept to a minimum. For example, if the starting materials and the reagents have to be dissolved to bring them into solution and reaction mixture has to be heated to complete the reaction, in such case the time required for completion of the reaction should be minimum so that the energy requirements is least. Use of a catalyst has the added advantage of reducing the energy requirement for a reaction.

In case the final product is impure, it has to be purified by distillation, re-crystallisation etc. for which energy is required. The process should be designed, as far as possible, in such a way that there is no need for separation or purification, so that the energy requirements are minimum.

Energy can be supplied to a reaction by photochemical means, microwaves or sonication. These procedures take much less time with minimum energy.

7.9.6 Selection of Appropriate Solvent

The solvent selected should not cause any health hazard or any environmental pollution. Use of supercritical CO_2 should be explored. If possible, the reaction should be carried out in aqueous phase or in the solid state without the use of the solvent.

In case of solvents, the volatility is one of the problems, which may affect human health and the environment. A novel class of solvents, generally referred to as green solvents, is that of versatile ionic liquids. These are non-volatile and have excellent solvent properties.

7.9.7 Selection of Starting Materials

As far as possible, the starting materials should be obtainable from renewable sources. Thus, petrochemicals are considered to be obtained from non-renewable sources in the sense that it takes million of years for the formation of petroleum oil from vegetable and animal remains. The staring materials, which can be obtained from agricultural or biological products, are referred to as renewable staring materials. The main problem with these materials is that such products cannot be obtained in a continuous supply due to factors like crop failure etc.

Gases like carbon dioxide (generated from natural sources or synthetic routes) and methane (obntained from natural sources like natural gas or marsh gas) are available in abundance. These are considered as renewable staring materials.

7.9.8 Use of Protecting Groups

In organic synthesis, if there are two reactive groups and we want to use only one of the groups, the other group has to be protected. After the reaction is completed, the protecting group is finally removed. One such example is given below.

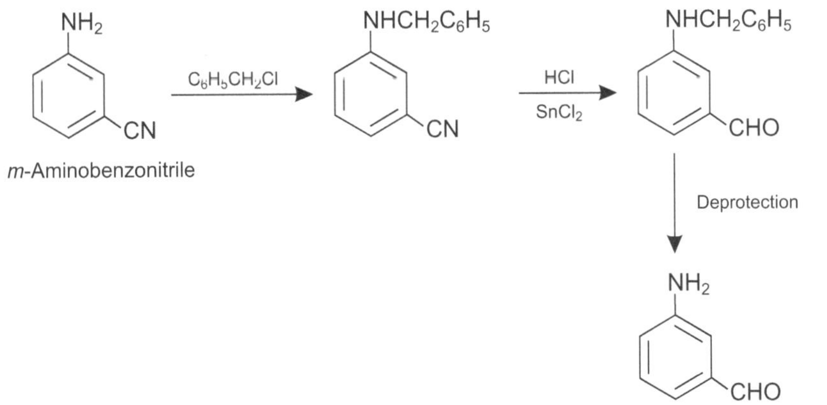

We see that the protecting groups are needed to solve a chemoselectivity problem. The protecting reagent should be added in stoichiometric amounts only and removed after the reaction is complete. These protecting groups are not incorporated into the final product and hence, their use make a reaction less atom-economical. As far possible, the use of protecting groups should be avoided.

7.9.9 Use of Catalyst

If possible, a catalyst should be used to facilitate transformations at comparatively low temperatures. The catalyst is not incorporated in the final product and can be reused. Some advantages of the use of catalyst are as follows:

(*i*) Better Yield

Hydrogenation or reduction of olefins in presence of nickel catalyst

$$\underset{\text{Propene}}{CH_3CH{=}CH_2} + H_2 \xrightarrow{\text{Ni}} \underset{\text{Propane (100\% yield)}}{CH_3CH_2CH_3}$$

Preparation of cyanides

$$\underset{\text{Benzyl chloride}}{C_6H_5CH_2Cl} + \text{aq. KCN} \xrightarrow[\text{Transfer catalyst}]{\text{Phase}} \underset{\text{Benzyl cyanide (90\% yield)}}{C_6H_5CH_2CN}$$

Oxidation of side chain

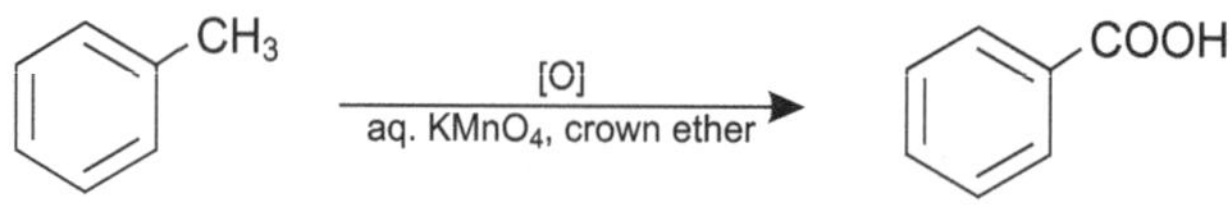

Toluene Benzoic acid (85% yield)

(*ii*) The reaction proceeds even in those cases where no reaction is normally possible without the use of catalyst.

$$\underset{\text{Acetylene}}{HC{\equiv}CH} + H_2O \xrightarrow[H_2SO_4]{HgSO_4} \underset{\text{Acetaldehyde}}{CH_3CHO}$$

$$\underset{\text{Propyne}}{H_3CC{\equiv}CH} + CO + CH_3OH \xrightarrow{\text{Pd}} H_3C{-}\underset{\underset{CH_2}{\|}}{C}{-}\overset{\overset{O}{\|}}{C}{-}OCH_3$$

Methyl methacrylate
(shell corporation)

(*iii*) Enhancement of selectivity

$$\underset{\text{Propyne}}{H_3C{-}H{\equiv}CH} + H_2 \xrightarrow{\text{Pd-BaSO}_4} \underset{\text{Propene}}{CH_3CH{=}CH_2}$$

7.9.10 Products Designed should be Biodegradable

In case, the products obtained are not biodegradable, these keep on accumulating on the land or water bodies and are responsible for the environmental problems besides having other effects. Such a problem was encountered in insecticides and polymers. It is now well known that some of the insecticides (like DDT and organophosphates) are not biodegradable. These accumulate is many plants and get incorporated into food chain and finally into humans and animals causing extremely harmful effects. Also some of the insecticides are responsible for the decline of population of beneficial insects. In view of the above, it is very important that any product (e.g. insecticide) synthesized must be biodegradable. It is equally important that during biodegradation, the products obtained should not have any toxic effects or should not be harmful to human health. It is possible to have a molecule, which may possess functional groups that facilitates its biodegradation. The functional groups should be susceptible to hydrolysis, photolysis or other cleavage.

Rohm and Hass Company have developed some diacylhydrazines that are useful as insecticides and are biodegradable.

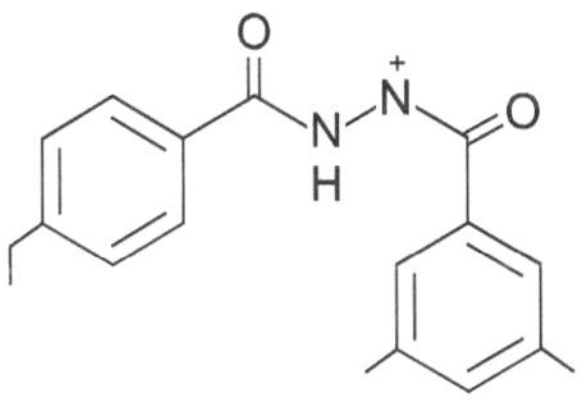

Tebutanozide

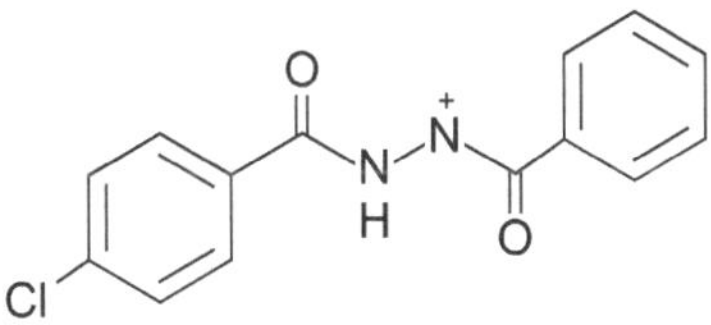

Halofenozide

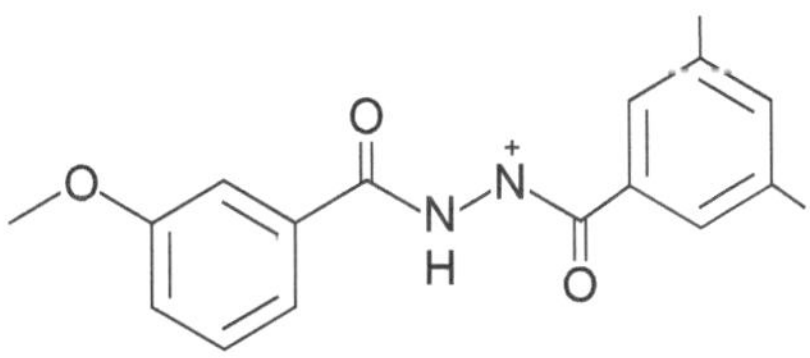

Methoxyfenozide

7.9.11 Designing Manufacturing Plants

It is extremely important to design manufacturing units in such a way that the possibility of accidents in minimum. A large number of accidents have occurred in industrial units. The gas tragedy in Bhopal (Dec. 1984) and several other places has resulted not only in loss of thousands of human lives but have also rendered numerous other people disabled for the rest of their lives. The manufacturing plants should be so designed that it eliminates the possibility of accidents during operation.

EXERCISE

1. What do you understand by Environmental Toxicology?
2. What toxic chemicals are present in the environment? Give their sources and toxic actions.
3. What do you understand by biodegradability? Give principles of biodegradation.
4. What type of transformations can be accomplished by microbes? Give examples.
5. What is the impact of toxic chemicals on enzymes?
6. Give biochemical effects of arsenic, cadmium, lead and mercury.
7. Write notes on:
   (*a*) Bhopal gas tragedy
   (*b*) Chernobyl disaster
   (*c*) Three Mile Island disaster
   (*d*) Minamata disaster.
8. What do you understand by green chemistry? How it will be helpful for having better industrial processes?

Chapter 8

Environmental Pollution and Health

8.1 INTRODUCTION

We know that since the industrial revolution, human society has made considerable progress in technological and industrial spheres. There is considerable growth in coal production, energy generation, petrochemicals, fertilisers, pesticides and chemical industry. These are likely to increase further in the years to come. All this has been accompanied by considerable degradation or pollution of the environment (i.e., air, water and land) due to large quantities of gas, liquid and solid wastes generated by industry and other human activities. Environmental pollution affects human health, directly or indirectly by undermining the life support systems of our biosphere.

8.2 EFFECT OF AIR POLLUTION ON HUMAN HEALTH

The air pollutants, as we know, are inhaled. This results in the attack of the pollutants on various parts of the respiratory system on their route particularly, to air sacks. Once the pollutants reach the body, they circulate throughout the body and reach to the different organs. The extent of damage depends upon a particular pollutant, its concentration in air and exposure time.

It has been established that air pollutants are linked with respiratory and some other diseases in human beings.

Sulphur dioxide is mainly responsible for lung diseases. It causes irritation to nose and mucous lining, shortness of breath, accumulation of tissue fluids,

Swelling (edema) and bronchospasm. Particulate matter potentiates the effect of SO_2. The gas settles on the particles and reaches to the deeper parts of respiratory system and show its deteriorative effects.

Vehicular exhaust cause eye and lung irritation. This is due to the presence of oxides of nitrogen. Inhalation of large amounts causes gum inflammation, internal bleeding, pneumonia and cancer.

Carbon monoxide is the most poisonous gas. High concentration of CO slow down physical and mental activities and may cause asphyxiation, heart and brain damages and ultimately death.

Ozone and peroxyacetyl nitrate (PAN) have marked biological activity. Even in small concentration, these cause irritation of eyes. Ozone causes cough, headache, weakness, difficulty in breathing, bleeding, dry-throat and constriction of respiratory passage and weakens lung tissues.

Effect of particulate matter depends on its nature and size. They cause irritation reactions. They are also constituents of photochemical smog, which is extremely harmful.

Fumes of toxic metals like lead, mercury, zinc and manganese as well as oxides of antimony, arsenic, copper etc are extremely harmful. Lead damages brain of young children and can cause death. In adults, it affects kidneys, blood, liver and the nervous system.

Dust of coal, asbestos, fire fibres of sugarcane, cotton etc cause repiratory problems. Long exposure may cause respiratory diseases known as **pneumoconiosis**.

Hydrocarbons like benzpyrene, benzene, chlorinated hydrocarbons, polybiphenyl chloride and many other compounds cause cancer.

Smoking also pollutes the air. Lung cancer, various respiratory and cardiac problems are linked with smoking. Non-smokers, or passive smokers as they are called, have as much risk as smoker.

The effect of air pollutants of health is summarised in Table 8.1

Table 8.1 Effect of air pollutants on health

Poluant	*Effects*	*Health effects*
Carbon dioxide (CO_2)	Responsible for greenhouse effect contributes to acid rain	
Carbon monoxide (CO)	Results in asphyxia by binding to haemoglobin. It is poisonous and is known as Silent killer	– Headache, fatigue, drowsiness – Reduces oxygen carrying capacity of blood – Health disorders – Effects CNS and causes brain and vision damage

(Table 8.1 Contd...)

Poluant	***Effects***	***Health effects***
Sulphur dioxide (SO_2)	– Affects lungs – Responsible for acid rain – Corrodes paints, metals – Damages vegetation – Responsible for the formation of smog	– Effects eyes, nose and mucous lining – Cough and choking – Oedema – Bronchitis, asthma – Aggravates heart and lung diseases
Sulphur trioxide (SO_3)	Precursor of acid rain Corrodes metals, paints Effects vegetation.	– Irritates eyes, nose and mucous lining – Cough, choking, oedama – Chronic asthma – Aggravates lung and heart diseases
Nitric oxide (NO)	Responsible for formation of secondary pollutants like nitrogen dioxide, nitric acid and peroxyacetyl nitrate (PAN)	– Supresses growth of plants – damages tissues – Irritation to eyes
Nitrogen dioxide (NO_2)	Harmful to vegetation	– Eye irritation, respiratory track – Reduces lung functioning
Suspended particulate matter (SPM)	Gets deposited on leaves and interferes in photosynthetic process Responsible for formation of smog.	– Irritation of throat, respiratory track – Effects lungs – Stressful effects on heart – Visibility is reduced
Dioxan	Carcinogenic effect	Responsible for causing cancer in humans
Asbestos	Causes lung diseases	– Asbestosis (lung cancer)
Coal dust	Effects lungs	Black Lung disease
Radionuclides	These are obtained by nuclear fission during the production of electricity or testing of nuclear devices	Effects different organs of the human system as given below Radium –226 Bones Radon –222 Lungs Iodine –131 Thyroid Krypton –85 Ovaries Cobalt –60 Liver Potassium –42 Muscles Sulphur –35 Skin Exposure to higher does of ionizing radiation causes cancer of various organs. It also damages brain, CNS, causes delirium, convulsions and death.
Noise	Cause noise pollution	Effects the hearing capacity of humans (for details, see chapter 7)

8.3 WATER POLLUTION AND HEALTH

The water pollutants include the following

(i) **Biological agents**: These include disease causing organisms like virus, bacteria and worms etc. They are commonly present in polluted water. The diseases caused by microorganisms in polluted water are discussed in Table 2.12.

(ii) **Chemical agents:** These include, inorganic chemicals (like nitrates, phosphates, toxic heavy metals etc), organic agents (like oil, gasoline, particles, dyes, paints and organic waste like domestic waste), animal manure etc.

The hazardous effects of some metals on human being are discussed in Table 2.13. The effects of toxic elements in water are discussed in Table 2.20.

Besides the above, the radioactive substancer are also chemical agents. The harmful effects of these are discussed in Table 8.1.

(iii) **Physical Agents:** These include suspended solids, temperature. The suspended solids can absorb metals and other toxins, and are passed into food chain. Temperature causes thermal pollution. This reduces the solubility of oxygen in water resulting in the death to marine organisms particularly fish.

An important water pollution is marine pollution, which is basically pollution by oil in seas and oceans. This type of pollution is harmful to the marine organisms.

Water pollution deteriorates the quality of water used for drinking, bathing or other activities. Water contaminated with sewage has foul smell and creates unhygienic conditions in the surrounds and affects our health. Drinking water polluted with sewage is a source of viruses, bacteria, protozoa and worms. Water borne infectious diseases like cholera, dysentery, typhoid, Jaundice and worm infection are still the major heath problems.

Excess quantity of **heavy metals,** such as mercury, lead, cadmium, zinc and copper are harmful to health.

Mercury compounds discharged in effluents are converted into highly toxic methyl mercury by bacterial action in the aquatic environment. In 1952, poisoning in Japan resulted in **'Minamata disease'** due to eating fish taken from mercury polluted Minamata Bay. In fish, methyl mercury can accumulate upto thousand time more than its concentration in the environment. In this disease, the victims developed numbness in the limbs, lips and tongue, and lost control. It also deafness, blurring of vision and mental derangement.

Cadmium poisoning caused another desease called Itai-Itai. It occured in Japan due to the consumption or rice affected with cadmium metal. The rice fields were irrigated with effluents released by zinc smelters. It is a painful disease of bone and results in the cancer of liver and lungs.

Excess to nitrate in drinking water causes **methemoglobinemia**. Nitrate converts haemoglobin to methemoglobin-oxidised (Fe^{3+}) form of haemoglobin which is non-functional. Nitrates can be fatal to human, especially infants under 3 months. Such infants are called **blue babies**.

Pesticides get accumulated in increasing quantities in organisms of higher trophic levels. Pesticides, such as DDT, aldrin, and dieldrin are suspected carcinogens. They cause tremors, convulsions causing damage to kidney. In India, the hazards of pesticides came to light by a case study of the people called **Endemic Familial Arthritis**, which affected poor Harigans of Malnad in Karnataka. The victims had eaten crabs picked from rice fields sprayed with pesticide and suffered pain in the hips and knee joints, and could not stand up as they were crippled.

Fluoride is known to prevent dental decay. However, if taken in excess causes mottled teeth, stiffening of joints and hardend bones called **skeletal fluorosis** or **knock knees** disease. The deformity of knees may lead to total disability to move. Excess of fluoride in water has been noticed in Tamil Nadu, Kerala, Gujarat, Rajasthan, Punjab, Haryana and Bihar.

8.4 EFFECT OF SOIL POLLUTION ON HUMAN HEALTH

Soil or land is polluted by the dumping of solid wastes generated in the household and manufacturing units. In addition to these domestic and industrial effluents disposed on land, run-off from agricultural fields and sewage also cause land pollution.

The severe health impact of the herbicide, 'Agent orange' (a mixture of two herbicides, 2, 4-D and 2, 4, 5-T) used by US forces during the Vietnam war by spraying the fields in order to destroy vegetation, which was used as a cover for the Vietnamese soldiers to hide is very well known. The people who were affected by these herbicides had off springs which were not only disabled, but their organs were also missing.

The impact of pesticides on human health are extremely severe. Acute effects occur relativately quickly on exposure to elevated levels of pesticides. These can encompass adverse effects ranging from headache, dizziness, irritability, and nausea to significant impairment of the nervous system and death. Chronic effects include the development of long term illnesses, such as cancer.

Besides the above, other sources of soil pollution include industrial pollution, urban and domestic wastes, radioactive pollutants, agrochemicals, plastic and polymers, chemicals and metallic pollutants, biological agents and acid rain (see also section 3.3.1)

The effects of soil pollutants like dioxan, asbestos, coal dust and radio nuclides are given in Table 8.1. The Effects of some metals on human are given in table 2.13 and the hazardous effects of toxic elements are discussed in Table 2.20

8.5 POLLUTANTS ENCOUNTERED AT HOMES (INDOOR POLLUTION)

In homes, the main causes of pollution are gas, kerosine, coal, wood and tobacco that release gases or particles into the air. Bio-fuels like fuel wood, crop residues, dried animal dung are used by more than 75 per cent of Indian household and

50 per cent of the world population for daily cooking. Bio-fuels are the major sources of indoor pollution. The gases and soot emitted on combustion contain six major pollutants, *vis.*, carbon monoxide, suspended particulate matter (SPM), polycyclic aromatic hydrocarbons, formaldehyde, oxides of nitrogen and sulphur, Due to these pollutants, there is higher prevalence of ailment in people in household using biomass fuels compared to people in the house hold using cleaner fuels. India at present bears the largest number of Indoor Air pollution (IAP) related health problems in the word. Approximately, 500,000 women and children die in India due to IAP related causes. The health effects of indoor air pollution are:

- Cough, breathlessness, chest illness chronic bronchitis
- Pulmonary hypertension and cardiac enlargement resulting in corpulmonale, decreased ventilatory capacity
- Certain type of cataracts leading to blindness
- Adverse pregnancy outcome–stillbirth, neonatal death or low birth weight
- Lower immune response in neonates
- Acute respiratory infection in children–cough, running nose, noisy respiration and wheezing
- Cancer and tuberculosis.

8.6 POLLUTANTS ENCOUNTERED AT WORK PLACE

By work place we mean, working in office, industrial units, agricultural jobs and persons employed in hospitals, and handling medical wastes. The pollutions associated with work place depends on the type of job one is doing.

Table below (8.2) lists the pollutants associated with working in office and using equipment like computers, cellular phones and xerox machines.

Table 8.2 Pollutants associated with use of office equipments

Equipment	*Type of pollution*	*Tentative concerns*
Computers	Emission of low intensity electromagnetic radiations from computer display systems	Excessive exposure leads to eye problems.
Cellular phones	Emission of microwaves while using.	Probably associated with brain disorder
Xerox machines	uv-Radiation and ozone exposure	Radiation and respiratory related hazard
Smoking	Carbon monoxide	Poisonous

Pollution encountered by people working in industries depend on the type of industry they are working and the nature of the job they are handling. Following table (8.3) gives some industries and specific pollutants encountered.

Table 8.3 Pollutants encountered working in industries

Industry	***Pollutants encountered***	***Effects***
Steel industry	Particles, CO_2, CO, heat, radiation etc.	Effects well known
Chemical industry	Various organic chemicals, hydrocarbons toxic gases	Effects well known
Cement industry	CO_2, cement dust	Pneumoconiosis
Coal mining	CO_2, coal dust	Pneumoconiosis
Silica mining	Silica, SiO_2	Silicosis, coughing, chest pain
Asbestos mining and factory	asbestos dust	Asbestosis, cancer of lungs
Cotton industry	Cotton dust	Byssinosis, lung diseases
Arc welding	UV and ozone exposure	Effects well known

EXERCISE

1. What is the effect of air pollutants on health.

2. Write notes on

   (*a*) Miniamata disease. How it is caused

   (*b*) Methemoglobinemia

   (*c*) Knock-knee disease

3. What is Indoor pollution. How it is caused and what is its effect human health.

4. What pollutants are encountered in work place and what are their effect on human health.

❑❑❑

Chapter 9

Energy and Environment

9.1 INTRODUCTION

The word 'Energy' is derived from the greek 'energon' meaning work. It is utilised by every one in the world. We use energy in our day-to-day life, in our homes, at work, in industry, and in agriculture. The prosperity of a country has come to be measured by the energy it consumes. In India, the important sources of energy are still cowdung and firewood, as these are freely and cheaply available. About 40 per cent of India's energy requirements comes from these two sources. The rest is from petroleum, coal, electricity, nuclear energy etc. There has been considerable emphasis on commercial production of energy in India, since Independence. As you have already learnt that it is better to utilise cowdung as a fertiliser and for the generation of biogas. The use of wood for producing energy should be discouraged because for wood, the forest have to be cut. It has already been stated that forests play an important role in the preservation of the ecosystem and so it is necessary to conserve our forests.

Each ecosystem is driven by energy. It has already been stated that we use energy in every part of our daily life. Science and technology have helped to convert energy from the naturally available forms into forms that can perform various tasks. Heat energy is converted into electrical energy and used to operate many modern machines and appliances. Electric energy is converted into heat for domestic purposes, such as cooking and heating. In fact, energy is a very vital part of the whole ecosystem.

We know that energy is transformed from one form to the other. We consume food to enable us to use our muscle energy for physical work. Physical work is employed to produce other forms of energy for human use. There is energy all around us with the sun being the primary source of all energy. Solar energy, as we know, converts carbon dioxide and water into organic compounds, which store the trapped solar energy, in the form of chemical energy. This kind of chemical energy is transformed into coal, oil and natural gas, which is again converted into heat energy to be used for a variety of purposes. The vast expension of industry and agriculture sectors depend largely on available energy sources. About 88 per cent of world's present energy production comes from coal, oil and natural gas. Apart from depletion of these resources, there is added problem of pollution which is inevitable.

Other sources of energy are hydro energy and nuclear energy which contribute to the balance of commercial energy.

9.2 SOURCES OF ENERGY

There are two types of sources of energy. These are **renewable** energy sources and **non-renewable** energy sources. The renewable energy sources also known as non-conventional energy sources include solar energy, hydro energy, tidal energy, wind energy and biomass energy etc. On the other hand, the non-renewable energy sources are known as conventional energy sources and include fossil fuels, coal reserves and petroleum and natural gas. Besides these two types of energy resources, we have another very important resource in the from of radioactive minerals, particularly uranium (^{235}U). These are used to produce the so called nuclear energy or atomic energy. The amount of ^{235}U available in nature is expected to last for many years.

The demand of energy approximately doubles every 14 years and as already stated is taken as one of the indicators of development of a country. India, with 16 per cent of the world's population consume only about 2 per cent of the total energy produced in the world, in comparison to USA which has 6.25 per cent of the world's population and utilizes 33 per cent of the energy produced. This gives us an idea of the low level of our development and should be an incentive for generation of more energy, so that we can meet our growing demands of energy. Even today, about 80 per cent of our population continues to depend on fuel wood, dung and agricultural wastes. It is well-known that non-renewable sources of energy, such as fossil fuels, coal and petroleum, are not going to last for long. Forests are also being depleted at a fast rate due to indiscriminate felling of trees. In view of all this, it has become necessary to think of alternative, non-conventional sources of energy.

Energy needs in India are met by harnessing three categories of energy sources as shown as follows:

- Energy
  - Non-conventional or Renewable Energy Sources
    - Solar energy
    - Hydro energy
    - Tidal energy
    - Wind energy
    - Biomass energy
  - Conventional or Non-renewable
    - Coal
    - Petroleum
    - Natural gas
  - Nuclear or Atomic Energy

9.2.1 Renewable Energy Sources

In nature, we have plentiful supply of energy fuels which can be renewed. These are renewable energy sources as they can be reproduced and can be harvested continuously. Solar energy, wind energy, water energy in the form of hydro electric and tidal energy are being harnessed more and more these days. Renewable sources also include firewood or fuel-wood, obtained from forests, agricultural wastes like biogas, animal dung etc. However, fire wood cannot be harnessed in short time intervals as is the case of other. The simple reason is that once the trees are cut, it take a long time to grow. However, with proper planning to the forest system more or less continuous supply of fuel wood (in reasonable quantities) can be maintained.

9.2.1.1 Solar Energy

Solar energy is the most readily available abundant source. It is free as it does not belong to anybody. It is also non-polluting. Solar energy has the greatest potential of all sources of renewable energy. Only if a small amount of this form of energy could be used, it will be one of the most important supplies of energy especially when other sources in the country are being rapidly depleted.

It is estimated that the sun gives us 1000 times more power than we need. If we can use even 5 per cent of this energy, it will be 50 times what the world will require.

Solar energy is being used in many part of India for cooking by using the so called 'solar cookers' (Fig. 9.1).

Fig. 9.1 Solar-cooker

The solar energy can be used directly to give us hot water during winter. This procedure is used in hotels and industrial establishments to get hot water (Fig. 9.2)

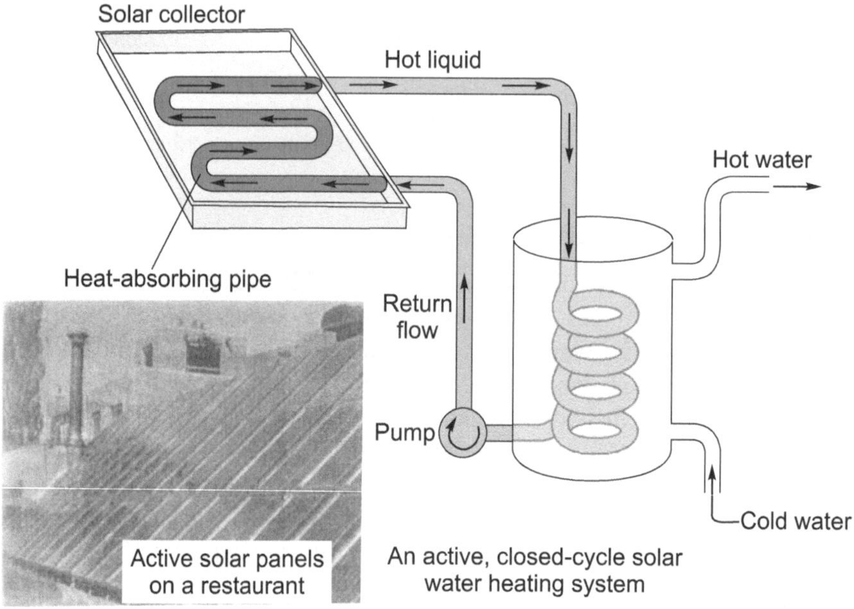

Fig. 9.2 Solar-water heating system

Solar run refrigerators have been developed for rural areas. These keep vegetables, fruits fresh for a longer period. This procedure has great potential for using solar energy for cold storages. It can also be used for room heating in colder regions. These days, high multistorey building use more glass in buildings to trap sun's heat. This has lot of potential even in less sunny climates.

Solar energy can be used with the help of 'photo voltaic cells' for producing electricity for driving vehicles and for illumination. In solar facility, hundreds of mirrors are used to focus sun's energy on a central collector; which in turn attains extremely high temperatures; these are useful for the production of electricity or industrial applications.

Solar Energy flow through ecosystem

The basic source of energy is from sunlight. The green plants through the process of photosynthesis, convert radiant energy of the sun into chemical energy. The portion of the radiant energy absorbed by the green plants and stored in the form of chemical energy is the food energy. It should be understood that out of enormous amount of energy continuously radiated by the sun is reflected or refracted back by the atmosphere surface of the earth and other objects. Only about one per cent of the solar energy received by the plant is used through the process of photosynthesis.

The plants can grow by converting the sun energy directly into their tissues. The plants are used by herbivores (the primary consumers) as food which give them energy. A large part of this energy is used up for the metabolic functions of these animals, such as breathing, digesting food, supporting growth of tissues, maintaining blood flow and body temperature. Energy is also used for activities, such as looking for food, finding shelter, breeding and rearing the young ones.

When the primary consumers (the herbivores), in turn may be eaten by a carnivore (the secondary and tertiary consumer), the energy is transferred from a primary consumers to a secondary or a tertiary consumer. In this way, transfer of energy takes place from one level to the other. The process of transfer of energy in a food chain is called **energy flow**.

About 10 per cent of the energy available in one tropic level is transferred to the next tropic level. If the secondary level receives 10 per cent of the energy of the producer, the tertiary level captures 10 per cent of this energy. This is referred to as 10 per cent law. Due to this 10 per cent law we find less number of carnivores as compared to herbivores. Of course, it is much easier to find plants in the ecosystem The energy loss at each trophic level is due to several factors as:

- Organisms use energy in respiration and other metabolic activities
- Some organisms are not consumed by the next trophical level and therefore, get decayed and decomposed.
- Some of the food energy eaten by the organisms is not assimilated and passed out of the body.

The fig. 9.3 below shows that the biological pyramid of an ecosystem depicting the number of organisms and energy available at each tropic level.

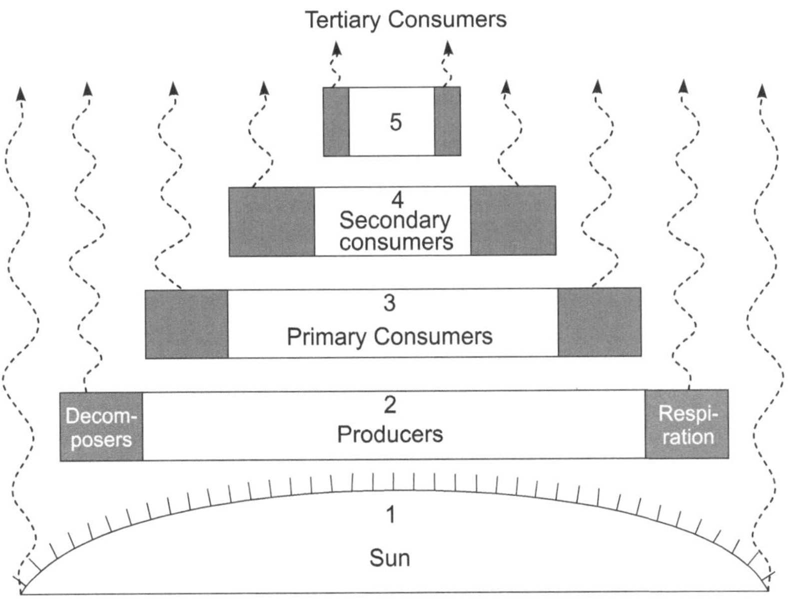

Fig. 9.3 Biological Pyramid in an ecosystem.

In the above biological pyramid note the following points:

- Some part of solar energy dissipates as it passes through the ecosystem.
- A small portion of the solar energy is captured by the plants (producer).
- At each trophic level, a part of the energy is captured from previous tropic level. This is because at each energy level some energy is used for metabolic purposes and also some goes to decomposers. Due to this type of energy transfer not much energy is left after 4 or 5 levels
- The number of organisms is also reduced as we go up the tropic levels.

The flow of energy from the sun in the ecosystem can be represented as shown below (Fig. 9.4).

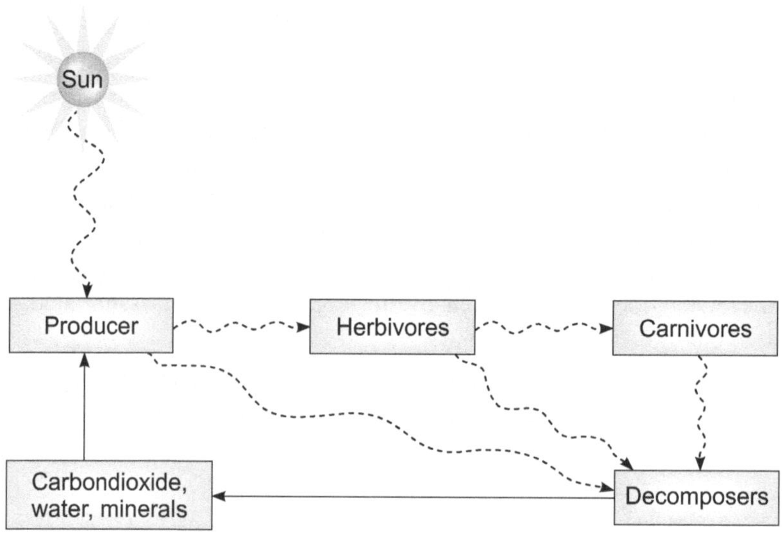

Fig. 9.4 Energy movement in an ecosystem. Note that the movement of energy is unidirectional

9.2.1.2 Hydro Energy

Hydroenergy is the cheapest, and cleaner and hence the best source of energy. Water power is developed by allowing water to fall under force of gravity and is used almost exclusively for generation of electric power. Potential energy of water in converted into mechanical energy by using prime movers known as hydraulic turbines. Hydroenergy is quite cheap in places where water is available in abundance (see Fig. 9.5). Although, the capital cost of hydroelectric power plants is higher when compared to other types of power plants, their operating costs are quite low, as no fuel is required.

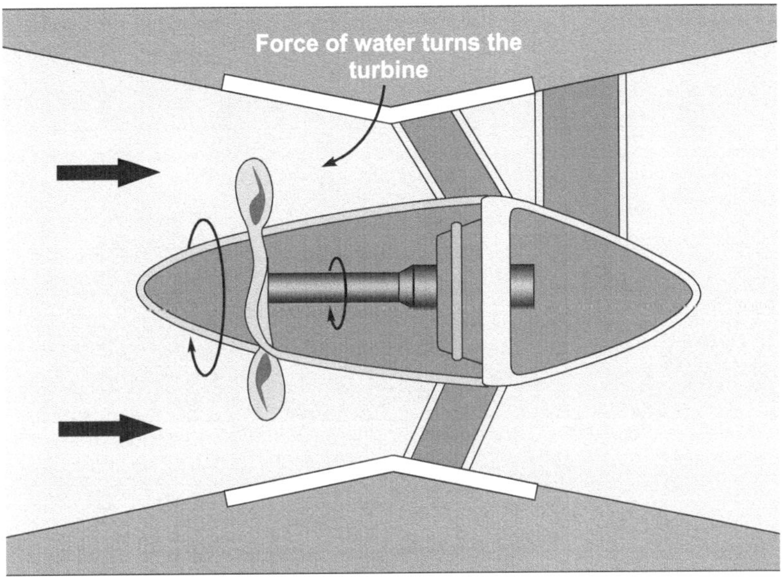

Fig. 9.5 A turbine to generate hydro-electricity

However, obtaining electricity from mega dams has given rise to many controversies in recent times, small hydro plants are emerging viable alternatives. These plants serve the energy needs of remote and rural area where the grid supply is not available.

A natural or artifical water fall is made to turn a modern type of pedal wheel, called a turbine, which upon rotation generates electricity.

9.2.1.3 Tidal Energy

Tides in the sea are the result of universal gravitational effect of heavenly bodies like the sun and moon on earth. Tides can be used to produce electric power, which is known as tidal power. The use of tides for generation of electric power is practical only in a few favourably situated sites where the geography of an inlet or bay favours the construction of a hydroelectric plant.

A tidal plant (Fig. 9.6) has been completed in France with a rating of 240 MW in 1960. In India, the first tidal energy project with a capacity of 150 MW has been set up at Vizhinjan near Trivandrum. A major tidal wave power project costing Rs 5000 crores is proposed to be set up in the Gulf of Kutch in Gujarat. Other suggested possible sites for tidal power plants are Gulf of Cambay and Sunderbans areas of West Bengal.

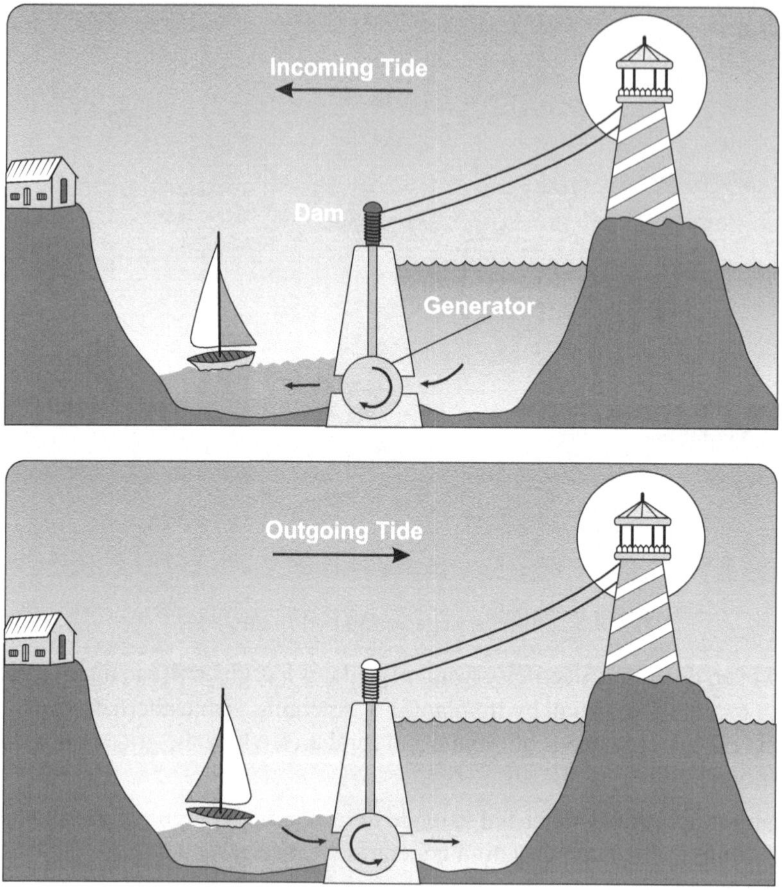

Fig. 9.6 Tidal power station

Both incoming and outgoing tides are held back by a dam. The difference in water levels generates electricity in both directions as water runs through reversible turbogenerators

9.2.1.4 Wind Energy

Wind energy has been used for hundreds of years for sailing, grinding grain and for irrigation. Energy in wind can be economically used for generation of electrical energy and has great potential as a good source. The energy available in the winds over the earth's surface is estimated to be 1.6×10^7 MW, which is equivalent to the present energy consumption in the world.

Wind is used to run a wind mill, which in turn drives a generator to produce electricity (Fig. 9.7).

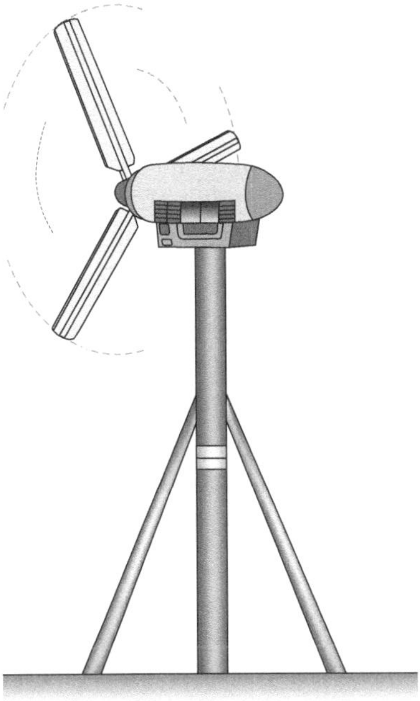

Fig. 9.7 Wind mill

Wind can also be used to provide mechanical power, such as pumping water. Wind mills for water pumping have been installed in many countries particularly, in the rural areas.

In India, generally wind speeds obtainable are in the lower ranges. Attempts are being made to develop low cost, low speed wind mills for irrigation of small and marginal farms, and for providing drinking water in rural areas. High wind speeds are obtainable in coastal areas of Sourashtra, Western Rajasthan and some parts of central India. In these areas, medium and large sized windmills could be used for generation of electricity.

Wind energy is an eco-friendly power source. It is a renewable source of energy, is non-polluting and does not need any fuel consumption. It works up to a few kilowatt system and is less costly on a small scale.

9.2.1.5 Biomass Energy

Biomass is a renewable energy source, which is derived from plant resources, animal waste and the waste of various human activities. It is also derived from the by products of the timber industry, agricultural crop, raw material from the forests, major parts of house hold wastes and wood. Biomass is an important source of energy and the most important fuel world wide after coal, oil and natural gas. The importance of biomass is that it can be used to generate electricity with the same equipment of power plant that are now burning fossil fuels.

Biomass fuels used in India account for about one third of the total fuel used in the country. Over 90 per cent of the rural households and about 15 per cent of the urban household use biomass fuels (e.g., wood, cowdung cakes, crop residues, sawdust etc.) However, the inefficient burning of such fuels in traditional *chulhas* causes serious problem of indoor air pollution and consequent health hazards. Also, the unsustainable level of consumption of fuel wood leads to deforestation which is responsible for environmental degradation.

Biomass resources fall into the following three catogeries:

(*i*) Biomass as traditional solid mass like wood and agricultural residue. Such a biomass is burned directly and energy obtained

(*ii*) Biomass in non-traditional form. Such biomass is converted into ethanol and methanol which are used as liquid fuels in engines.

(*iii*) Biomass for domestic use. This is obtained by fermenting the biomass anaerobically to obtain gaseous fuel called biogas. This is the major source of domestic energy.

Biogas: Biogas, as you know is a source of energy for cooking. Using a simple process, cattle dung is used to produce a gas (called biogas) that contains 55-70 per cent of inflammable methane gas. It is a clear and efficient fuel for rural areas. Water weeds like water hyacinth, water lettuce, salvinia, hydrilla, duck weeds and algae are found to be useful supplement to cattle dung. Biogas can also be used to generate steam, which in turn may be used for running engines or machines in factories or for running turbines to generate electricity. Biogas plants (see Fig. 9.8) can supply the needs of a number of families or even small villages. The residual dung or the digested slurry left after generating biogas can be used as manure for agricultural purposes. This is an economical way of obtaining energy from organic wastes. In India, great efforts are being made to install thousands of biogas plants in rural areas.

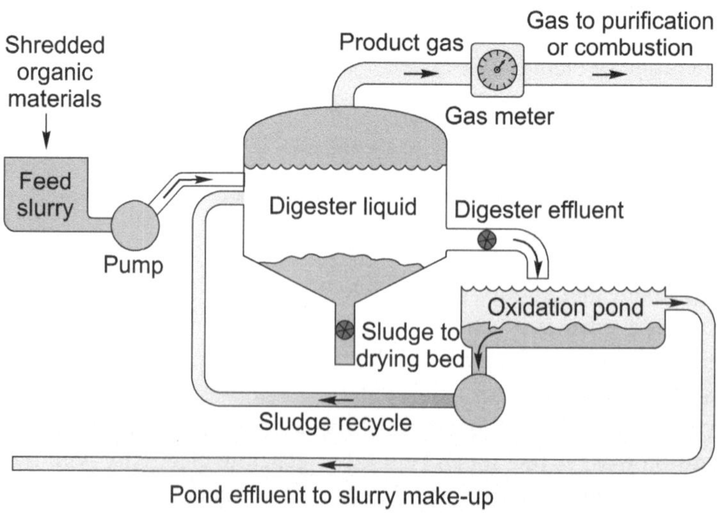

Fig. 9.8 Biogas production by anaerobic fermentation

Status of power generation in India from renewable resources

The following table gives the status of renewable energy resource use in India.

Table 9.1 Renewable Energy Resource Potential and Current status

Resource	*Gross Potential*	*Achieved*
Wind	45000 MW	1628 MW
Solar Thermal	35 MW/sq km	—
Solar Photovoltaic	22 MW/sq km	2.24 MW
Biomass	3500 MW	381 MW
Small Hydro	15000 MW	1663 MW

Source: Ministry of non-conventional Energy Sources (MNES) Govt. of India

As seen from the above data, we have still a long way to go.

9.2.1.4 Geothermal Energy

It is the heat energy, which is present deep down in rocks (several miles below the earth surface). A stage is reached, where temperatures are extremely high in molten rocks called magma (temperature more than 1000°C). This earths energy can be converted into heat and electricity. At least three types of power plants are available, which operate using geothermal energy.

Geothermal energy comes out of earth in the form of column of water and steam, called gyser. There are a number of places in the surface of earth, usually in the areas of volcanic and earthquake activity, where immense heat comes out of the surface in the form of water and steam at temperature in the region 300°C. The source of this water and steam is ground water which has found its way from the earth's surface down through porous rocks. This water gets heated by the rock and comes out in the form of gyser. This steam is used for heating homes and generating power. A geothermal power plant is establesed in Northern California, Italy, Japan, USSR, China.

9.2.2 Non-Renewable Energy Sources

Non-renewable energy sources are those like coal, petroleum and natural gas. These are also called *fossil fuels*. These sources are available in limited amounts and develop over very-very long periods of time. With rapid increase in human needs, population growth and rapid industrialisation, these sources are likely to be exhausted before long. In view of this, efforts must be made to develop renewable sources of energy.

9.2.2.1 Coal

Coal reserves have been derived from prehistoric forests by a process of decay and transformation. Coal was formed from the remains of trees and ferns which grew in swamps some 500 million years ago. Initially, bacterial and chemical action on such plant debris produced peat (dead plant material) as an intermediate product. Peat is composed of dead leaves, stems and roots of plants is mainly cellulose. As a result of high pressure and temperature during geological changes under earth's surface, peat was transformed into coal. Following are the sequences of changes which take place for the formation of coal

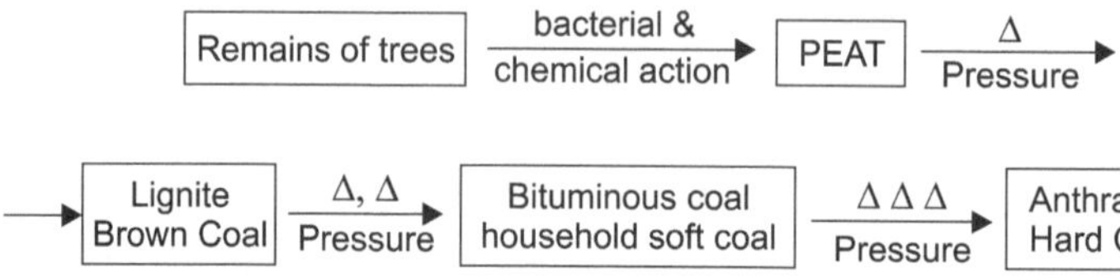

Coal has been a major contributor to world's energy needs. The early part of the twentieth century was called the "coal-age" as almost 80 per cent of energy requirement of the world was supplied by this source. In India, our coal consumption provide about 30 per cent of our energy needs.

Composition of Coal: Coal is a mixture of substances containing volatile matter that can be easily vapourised. It has a varying amount of fixed carbon, which is the solid part that burns after the volatile matter and moisture have been driven off.

Types of Coal: Coal is classified into three main types, *viz.*, lignite, bituminous and anthracite. As already stated, these are derived in different stages during the formation of coal. The percentage of carbon in peat, lignite, bituminous and anthracite is approximately 54, 64, 84, and 93 per cent, respectively. Besides carbon, these varities of coal also contain nitrogen (about 1–2%) and sulphur (1 – 3.5%).

Pollution Caused by Coal: In India, coal reserves are substantial. It has about 5.7 per cent of the reserves of coal in the world. The major consumers of coal are the power sector, steel, cement, railways, fertilizers and the house hold sector. Direct combustion is the cause of major air pollution. Gases, such as oxides of carbon, nitrogen and sulphur are toxic. Apart from general environmental effects on the health of the community, those working directly in such plants are exposed to severe health problems including cancer.

Carbonisation of cocal (heating about 900 – 1000° C) given (besides aromatic hydrocarbons, ammonia liquor and coal tar) coke oven gas, which contains methane and higher alkanes (about 50%) along with hydrogen (10%), carbon monoxide (5%) and carbon dioxide (9%) etc. It was used in India for heating purposes (domestic consumption). Earlier, the cities in Mumbai and Calcutta had gas plants for supplying 'coal gas' to houses. However, this practice has been discontinued and now LPG in used for this purpose.

The residue left after carbonisation of coal, is the well known 'Coke' which is used for industrial heating including power generation.

In India, coal is mostly found in Jharkhand, West Bengal, Orissa and Madhya Pradesh. The main coal mines are in Jharia and Bokaro of Jharkhand, and Raniganj in West Bengal.

9.2.2.2 Petroleum

Petroleum is today, what coal was earlier. It is the major contributor to the world's energy requirements. Petroleum is an oily, thick and usually dark coloured liquid. It is found deep below earth's crust trapped within rock structure. The origin of the word petroleum is from the latin word's ptra (meaning rock) and oleum (meaning oil). The biggest oil producing countries in the world is USA. Other major oil producing countries are Russia, Venenzuela, Iran, Gulf countries, Rumania and Mexicco. In India, petroleum industry has made tremendous headway only after independence. In the last 30-35 years, petroleum production in India has increased almost 20 times. Even with this increased production only about 50 per cent of our needs are met.

Oil fields in India: In India, petroleum is produced in Assam (Makum, Rudrusagar and Lakuwa), Gujarat (Ankaleshwar and Kalol) and Bombay High (about 150 Km offshore Bombay and about 100 m below the sea bed).

Formation of Petroleum: Petroleum is believed to have been formed by the decay of remains of ancient microscopic marine plants and animals under the layers of rock and clay over millions of years.

Composition of Petroleum: Petroleum is a mixture of hydrocarbons in varying proportions. The actual composition of petroleum oil varies with the place of origin. Besides hydrocarbons, petroleum contains organic compounds containing nitrogen and sulphur. In fact, a large number of organic compounds are isolated from petroleum oil. These organic compounds are known as Petro chemicals.

Drilling for oil: Petroleum occurs at a depth of five to six hundred kilo meters. So it is obtained by drilling process. The first step, of course, is the location of the oil fields. The next step of drilling is carried out by an efficient rotary drill, which bites into the rock. After the drill strikes the oil, the underground pressure of the gas (known as petroleum gas or natural gas) forces the oil to the surface. The flow of oil from such wells usually continues for years. As the pressure to gas coming out subsides or when there is no gas, air pressure is applied to lift the oil from the well (see Figs.9.9 and 9.10).

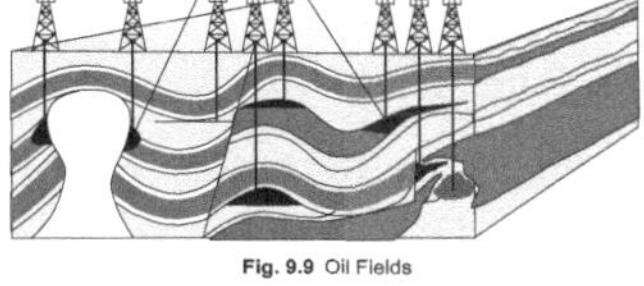

Fig. 9.9 Oil Fields

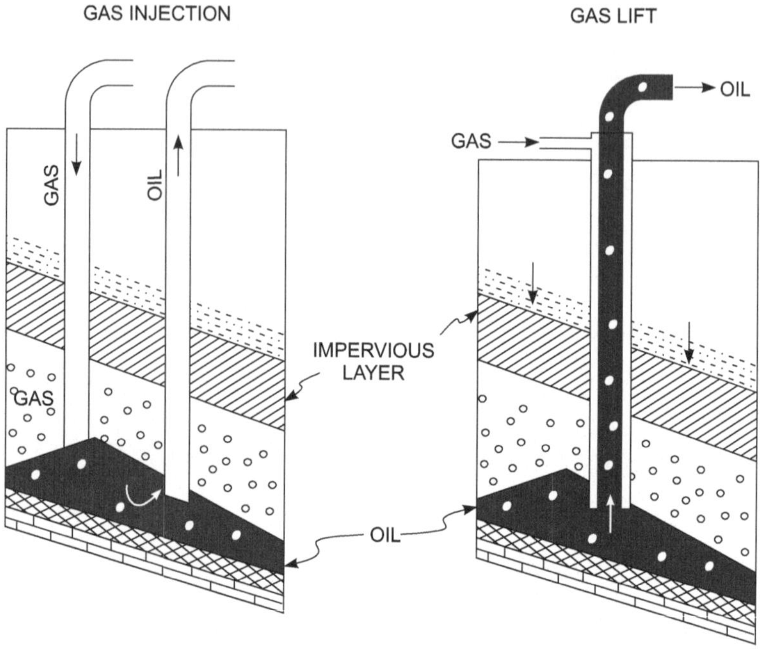

Fig. 9.10 Getting Petroleum from oil well

These days petroleum is also mined by drilling into the seabed. In India, this procedure is used for getting oil from Bombay High (Fig. 9.11)

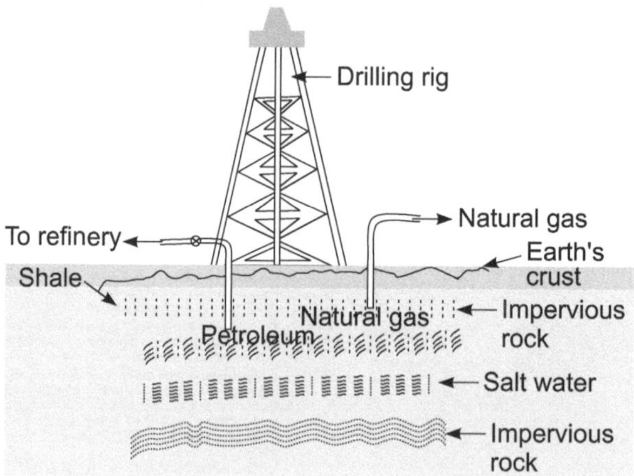

Fig. 9.11 Petroleum from deep inside the sea

Petroleum refining: The petroleum obtained by drilling process is known as crude oil or petroleum oil. It is obtained as dark coloured liquid, and is obtained mixed up with sea water and fine particles of slit (if the oil is drilled by off shore drilling). The crude oil is separated into a number of fractions by fractional distillation.

Fractional distillation of crude oil: The crude oil, as alredy stated chiefly consists of a mixture of hydrocarbons (and also compounds containing nitrogen and sulphur) boils in wide range (0° to 400°C) and is not suitable for any purpose. So, it is subjected to fractional distillation in order to separate it into a number of fractions. Each of these fractions has a technical name and is used in industry. The process of division of petroleum into fractions with different boiling ranges is termed **Refining** (Fig. 9.12)

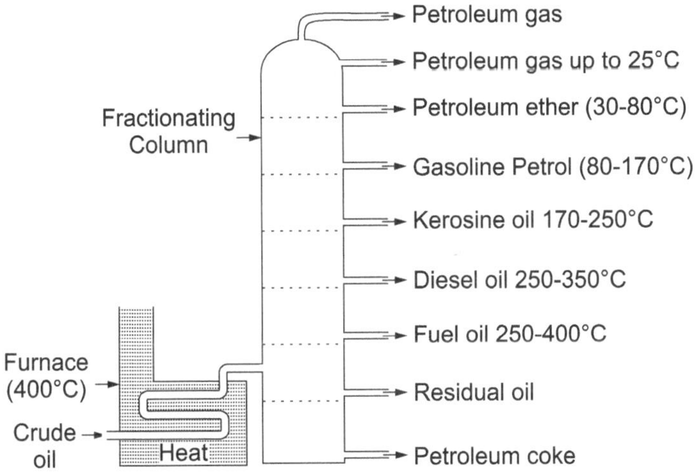

Fig. 9.12 Flow sheet of petroleum refining

The distillation of petroleum is carried in a tubular furnace which is a tall steel fractionating column. The crude oil is pumped continuously through heated pipes (400 °C) and fed into the fractionating column. The vapours of the oil as they rise up the fractionating column become cooler and condense at the selves at various heights. The highest boiling fraction condenses at the bottom and the lowest boiling fraction at the top. The uncondenses gases escape along with gasoline vapours at the head of the column. Outlets are provided in the side of the column to withdraw a number of fractions.

Products of fractional distillation: Various fractions obtained by fractional distillation of crude petroleum oil from top to the bottom of the column arc.

Petroleum gas: The gaseous fraction collected up to 25°C constitute the petroleum gas. It contains mostly methane, ethane, propane and butane. Butane is the major constituent. It is used as a portable fuel. After liquefaction, it is filled in cylinders and used as LPG (Liquefied petroleum gas). It is mostly used as domestic fuel. These days it is also used as a substitution of petrol for running vehicles (cars, buses, etc). This is a clean fuel and avoids any atmospheric pollution.

Petroleum ether: The fraction collected between 30–80°C is petroleum ether. It is mostly used as solvent for extraction of oils and fats in industry.

Gasoline petrol: The fraction collected between 80–170°C is the well-known petrol or gasoline. It is obtained as a colourless liquid and contains hydrocarbons (C_5 to C_9), both open chain and branched chain hydrocarbons.

It is used as motor fuel, aviation spirit, also used for dry cleaning.

Kerosene: The fraction boiling between 170°C and 250 °C is the well known kerosene. It is also obtained as a colourless liquid and contains compounds (hydrocarbons) containing 10 to 12 carbon atoms. It is used as a fuel for stoves, for lamps as illuminant, and fuel for jet aircraft engines.

Diesel oil: Diesel oil is collected in the range 250° to 350°C. It is a brownish liquid and contains hydrocarbons containing 13 to 15 carbon atoms. It is used as a fuel for high speed diesel engines in trains, buses and cars.

Residual oils: The residual oil left after 400 °C are used to obtain.

(*i*) **Lubricating oil:** Containing hydrocarbons C_{15} to C_{18}. Mostly used for lubricating machines

(*ii*) **Paraffin wax:** It contains hydrocarbons containing 18 to 30 carbon atoms. It is used in ointments, making candles and wax paper.

(*iii*) **Asphalt:** A black sticky substance, used for surfacing roads.

9.2.2.3 Natural Gas

The gas which flows out naturally from oil wells is referred to as **Natural gas.** In many parts of the world, large quantities of natural gas is found in association with petroleum. It is also found dissolved in natural petroleum and is given out during fractional distillation of petroleum oil. Some wells produce only the gas and no petroleum oil. The natural gas usually contains methane (95%) along with ethane, propane and butane (about 3.6%) along with carbon dioxide (0.48%) and nitrogen (about 1–2%). The proportion of ingredients of natural gas varies with the source. C_2 to C_4 gases are present to the extent of about 20 per cent in petroleum field gases.

Natural gas like petroleum is also believed to have originated from the decomposition of organic matter trapped in the layers of sedimentary rocks.

Earlier, the natural gas was considered to be a hazard in the petroleum industry as it formed explosive mixture with air. So it was taken away from the source and burnt. The production and utilisation of the gas dates to the beginning of the 20th century is USA. It was only about 1950 that USSR and other countries also followed.

In India, the natural gas industry is in infancy and it largely derived the so called natural gas from the cracking units of petroleum refineries. It occurs in large amounts in India in the Bombay high, the Gulf of Cambay and Ankleshwar in Gujarat. New gas fields have been found in Tripura, Jaisalmer and in the Krishna-Godavari delta.

Natural gas as Energy Source: About 7 per cent of India's needs are met by natural gas, especially in power generation, fertilizer and petrochemical production. Natural gas can serve to reduce dependence on foreign oil. These days, Natural gas is also used in the form of compressed natural gas (CNG) for automobiles, particularly the buses.

Uses of Natural Gas: Natural gas is used for the manufacture of hydrogen and carbon black. For this purpose, the natural gas is pyrolysed (heated) at 1000 °C, when the methane present in natural gas is converted into hydrogen and carbon.

$$\underset{\text{methane}}{CH_4} \xrightarrow{1000°C} \underset{\text{carbon}}{C} + \underset{\text{hydrogen}}{2H_2}$$

This process is used for the manufacture of hydrogen gas. The hydrogen gas is made to combine inth nitrogen in presence of a catalyst to produce ammonia gas which is used for the manufacture of fertilizers.

The carbon obtained in the above reaction is of very fine variety and is called **carbon black**. It finds use in paints, black printing inks and in the production of automobile tyres.

The methane present in natural gas is also used for the manufacture of methyl chloride, chloroform, methylene chloride and carbon tetrachloride.

It is also used for the technical production of methyl alcohol, formaldehyde, acetylene and hydrogen cyanide.

Pollution caused by burning fossil fuels

Fuels like coal, wood, petrol, diesel and kerosine are used in homes, industry and transport. Though, these fuels supply the energy needs, but the products of their combustion cause severe air pollution. Following are given the pollutants, their source and harmful effects.

Pollutants	***Source***	***Effects***
Carbon monoxide[1]	Combustion of coal and petrol	Toxic gas
Carbon dioxide	Combustion of coal and petrol	Non-toxic gas. Excess CO_2 in the atmosphere is considered a pollutant, since it produces green house effect and is also responsible for global warming
Sulphur dioxide[2]	Combustion of coal and petrol	Produces suffocation and affects lungs
Oxides of nitrogen, Nitric oxide and Nitrogen dioxide	Combustion of coal and petrol	Both are toxic, but nitrogen dioxide is more harmful. Effects human beings and vegetables.
Particulate matter	Combustion of coal and petrol	Affects lungs.

Notes

1. Carbon monoxide is a deadly poisonous gas. On inhaling, it combines with the haemoglobin in blood and this blood cannot carry oxygen to other parts of the body. This causes respiratory problems, suffocation and even death.
2. Both sulphur dioxide and oxides of nitrogen react with air and moisture to produce sulphuric acid and nitric acid, respectively. These are responsible for acid rain.
3. The particulate matter get deposited in the lungs and are responsible for a number of health problems including TB.
4. Earlier lead tetra ethyl was used as additive to petrol to check the so called anti knocking. Its use has now been discontinued due to harmful effects of lead in the exhaust gases in the automobiles. The gasoline now used does not cause knocking. This is achieved by varying the composition of petrol so that it contains more branched chain hydrocarbons.

9.2.2.4 Nuclear Energy

Due to the problems associated with coal and petroleum, nuclear energy could be an important source of energy. It is projected as the fuel to replace fossil fuel energy. The resources for nuclear energy are believed to be about ten times more than fossil fuels. Nuclear energy is the heat energy released when tiny particular or atoms, are broken apart. Uranium, a mineral found in earths crust, is the main fuel used to produce nuclear energy. It is estimated that fission of one kilogram of ^{235}U will release energy equal to that obtained by burning three million tonnes of coal.

Uranium is found in India in Rajasthan, Bihar and Andhra Pradesh. For nuclear energy generation ^{235}U is used. It is a radio isotope of uranium. The element uranium occurs as uranite, a mineral form of uranium oxide.

Nuclear energy is already being used for various purposes like generation of electricity, propelling of ships and aircrafts as well as making of radioactive isotopes used in medical diagnosis and treatment of several diseases. However, there are drawbacks. The cost and maintenance of atomic power plants are astronomically high. Also, the effects of nuclear wastes on human life are extremely hazardous. The indiscriminate dumping of these wastes is a major problem in the world.

The advantage of nuclear energy is that there is no emission of oxides of carbon, nitrogen and sulphur. The major risk is from improper management of accidents to reactors. A number of nuclear episodes have been reported. These include chernobyl disaster (section 7.7.2), Three Mile island disaster (section 7.7.3) and the nuclear accidents in Japan, which occurred on 11th march, 2011 due to earthquake (9.0 magnitude on richter scale) followed by a 13 feet tsunami in few minutes; this was the strongest in the world since 130 years.

For more details about nuclear power see section 5.9 and 5.11.

Nuclear energy can be obtained either by **Nuclear Fission** or **Nuclear Fusion**. The Nuclear Fission (as already stated) is based on the fission of ^{235}U nucleir by neutrons

$$^{235}_{92}U + ^{1}_{0}n \longrightarrow ^{142}_{56}Ba + ^{93}_{36}Kr + 3\,^{1}_{0}n + 200\text{ MeV}$$

(for more details see section 5.9)

In nuclear fusion, two nuclei of lighter atoms (e.g., hydrogen) join or fuse to form necleus of heavier atom (e.g., helium). The process is accompanied by release of large amounts of energy.

EXERCISE

1. What do you understand by the term 'energy'.
2. What are the sources of energy. Which source in your opinion is the best. Give reasons.
3. How is coal formed in earth.
4. How is petroleum formed.
5. What is natural gas, biogas, coal gas. Give their uses.
6. In your opinion is nuclear energy is as best option. Give reasons.

Part–II

10. Experiments in Environmental Chemistry

Chapter **10**

Experiments in Environmental Chemistry

Experiment 1

Determination of pH of Soil

Determination of pH of soil is important to find out whether a particular soil is acidic or alkaline. In case, the soil is acidic or alkaline, suitable treatment can be done for reclaimation of soil. The is extremely helpful for cultivation of crops.

The pH can be measured by using a pH meter, which employs glass and calomel electrodes. Instead of two separate electrodes, a combined electrode can also be used. Different types of pH meters are available. These include a portable model (which is battery operated) or direct current operated laboratory model.

Procedure

(*a*) Calibration of pH Meter

(*i*) Follow the instruction given in the manual of the pH meter provided by the manufactures.

(*ii*) Switch on the pH meter by connecting it to power supply and let the instrument warm up. It takes about 2–3 minutes.

(*iii*) In a 100 mL beaker, take a standard buffer solution of pH7[1]. Note the temperature[2] of the solution and adjust the pH meter using the temperature compensation knob.

(*iv*) The electrodes[3] are taken out of the storage solution, rinsed with distilled water, dried with a soft tissue paper and placed in standard buffer solution of pH7.

(*v*) Gently stir the solution, taking care to ascertain that the electrode membrane is completely immersed in the solution and that the electrodes do not touch each other of the sides and bottom of the beaker.

(*vi*) The pH is measured and the pH meter adjusted as per instruction of the manufactures to pH 7 (if necessary). The pH meter is put on standing mode.

(*vii*) The electrodes are removed from the pH7 buffer, rinsed thoroughly with distilled water and dried (using a soft tissue paper) and immersed in second pH closer to the pH of the sample in another beaker. For acid range, the standard pH 4 buffer[1] or pH 9.18 buffer[1] (for basic range) is used. The solution is gently stirred and the pH reading is taken. This pH reading should be within 0.1 unit of the pH of the second buffer. In case, there is difference, the pH meter is set as per instruction of the manufactures at the pH of the buffer. The calibration exercise is repeated.

(*b*) Determination of the pH of the soil sample

(*i*) The soil sample[4] (about 25g) is taken in a 50 mL beaker and make its paste by adding appropriate amount of water (using a glass rod). The paste should be uniform. This can be checked by ascertaining that the paste does not flow to close a hole left by the rod. Free water should not be visible on the paste. In case there is free water, the paste is made again.

(*ii*) The electrodes are removed from the buffer solution (which was used for the calibration of pH), rinsed thoroughly with distilled water, dried (by using a soft tissue paper) and inserted carefully in the paste of the soil. The paste is gently stirred (using a glass rod) and the pH is measured when the reading is constant.

(*iii*) The electrodes are taken out, rinsed thoroughly with distilled water, dried (using soft tissue paper) and again inserted into the suspension of the paste (after stiring gently with a rod) and note the pH.

The procedure is repeated till at least two readings are constant.

Notes

1. The standard buffer solutions of pH 9.18, 7.0 and 4.0 are commercially available. Alternatively, these can be prepared as follows.

   For making pH 9.18 buffer solution, 3.800g of $Na_2B_4O_7$. 10 H_2O is dissolved in distilled water in a one litre volumatric flask and brought to volume mark with distilled water.

   For making pH 7.0 buffer solution, 3.3.91g citric acid ($C_6H_2O_7$) and 23.3844g of disodium phosphate (Na_2HPO_4, $12H_2O$) is dissolved in distilled water and made up the volume to one litre in the volumatric flask.

   For making pH 4.0 buffer solution, 11.806 g citric acid is dissolved in distilled water and volume made up to one litre in the volumatric flask.

The standard pH solutions can be stored for 4 weeks in polyethylene bottles.

2. The pH is dependent on the temperature and so it is necessary to set the pH meter at that temperature before taking the readings.
3. In place of two electrodes (glass and calomel electrodes), a single combined electrode can also be used.
4. The sample of the soil in a lawn should be taken form the top 5 or 8 cm. of the soil. In case of garden or farm crops, soil sample be taken from the top 14 to 18 cm of soil. In case the area is large, soil samples should be taken from a number of locations and mixed together.

   The soil samples should be taken in sample tubes or in a plastic bag using a spoon or knife. The soil should be free of stones, leaves etc. The can be ascertained by passing soil through a 2mm sieve. Only the soil which passes through the sieve be used.

(*c*) Observations

1. Temperature of the standard buffer solution
2. Temperature of the soil sample
3. pH reading of the soil sample

S.No.	pH Reading
1	
2	
3	

(*d*) Result

The pH of the soil sample is ....

Experiment 2

Determination of pH of Water

Determination of pH of water is useful in the water softening and prevention of corrosion. In waste water, treatment involving biological process, the pH has to be controlled with in a favourable range.

As in the case of soil, the pH of water can also be measured using a pH meter.

Approximate pH of the water sample can also be determined using a universal indicator paper strip and comparison with the colour of the universal indicator.

Procedure

(*a*) Calibration of pH meter

The procedure followed is the same as in case of experiment 1.

(*b*) Determination of pH of water sample

(*i*) After calibration of pH meter, bring it to stand by mode.

(*ii*) The electrodes (or a single electrode) are removed from the second buffer solution, rinsed thoroughly with distilled water, dried gently (using a soft tissue paper) and immersed in the water sample[1] (taken in a beaker) for which pH is to be determined. The water is stirred (1–2 min) and pH measured when the reading is constant. If the temperature of water is different from the standard buffer solution, the temperature of the pH meter is adjusted using the temperature compensation knob.

Note and record the pH and temperature. Remove the electrodes and suspend in the standard buffer until the next measurement.

The procedure is repeated till at least two concordant readings are recorded.

Notes

1. The water sample should be free of any foreign matter and taken in a clean container of glass or plastic. As far as possible, the pH is determined as early as possible. In case water is taken from a tap, it is important to run water for few minutes before taking the sample.

(*c*) Observations

1. Temperature of buffer solution
2. Temperature of water sample
3. Measurement of pH of the water sample

S.No.	*pH Reading*
1	
2	
3	

(*d*) Result

The pH of the soil sample is ....

Experiment 3

Determination of Conductance of Water

Water conducts electricity due to the presence of dissolved salts. Thus, the measurement of the conductance of water, the approximate idea of the presence of soluble salts can be obtained. This procedure also gives an idea about the presence or absence of non-specific pollutants in fresh water. Even the salinity of seawater can be determined by this procedure. Also, the determination of the conduction of polluted water will give an idea of pollution of water.

The conductance of water is noted in conductivity terms, i.e., mS m^{-1} or μs cm^{-1}. The measurement of conductance of water indicates the concentration of total dissolved salts in water. It is most convenient to use the latest designs of conductometer, which directly displays the reading in conductivity terms i.e., mS m^{-1} or μs cm^{-1}.

Procedure

(*a*) Calibration of Conductometer

(*i*) Follow the instructions given in the manual of the conductomer supplied by the manufactures

(*ii*) Switch on the conductometer by connecting it to power supply and allow the instrument to warm up. It takes 2–3 minutes and set the mode selector on CAL

(*iii*) Using the range selector, set the conductometer to the desired setting, e.g., 200 or 220.

(*iv*) Using the sensitivity knob, set the meter reading to 1.0.

(*v*) Follow the manufacturers instructions manual and set the conductometer at 141.2 mS m^{-1}.

The conductometer in now calibrated.

(*b*) Measurement of Conductance of Water Sample

(*i*) The cell is rinsed with distilled water[1] and then with the water sample whose conductance is to be determined.

(*ii*) The sample water[2] is taken in a clean, dry beaker (100 mL capacity) and the conductivity cell dipped in it. Put the meter switch to READ. The conductance of water sample is read. Note the temperature to the nearest 0.1° C.

(*iii*) The conductometer is switched to CAL position and the cell is rinsed with distilled water and is kept in distilled water when not in use.

(*iv*) The conductance measurement is repeated to get at least two concordant readings.

(*v*) It is best to use a modern conductometer, which gives direct reading of conductivity in mS^{-1} or μS cm^{-1}.

Notes

1. Before use, distilled water should be boiled to minimise the CO_2 content. Electrical conductivity must be less than 0.01 mS m^{-1}.
2. The water sample is collected as in experiment 2.

(c) Result

The conductivity of water sample is .... μS cm^{-1} or mS m^{-1} (1μS cm^{-1} = μm hos cm^{-1})

Experiment 4

Determination of the Conductance of Soil

The conductance of soil depends on the presence of soluble salts present in it. In fact, the growth of plants depends on the presence of soluble salts in it. Thus, in order to determine the quality of soil, it is necessary to determine the level of soluble salts in it.

The conductance in soil is determined by extracting a soil sample with water and measuring the electrical conductivity of the water extract by means of a conductivity cell and meter.

Procedure

(a) Calibration of Conductometer

Follow the instructions given in the manual of the conductometer. Calibrate the conductometer as described in experiment no. 3.

(b) Measurement of the Conductance of Soil Sample.

(i) To 30g. of the soil sample[1] is added 60 mL of distilled water[2] in a 100 mL beaker. The suspension is stirred for about 45 minutes and then allowed to stand (about 25 minutes) so that the top layer is clear.

(ii) The temperature of the clear water is noted and conductance determined as described in experiment no.3.

(iii) It is best to use a modern conductometer, which directly gives the reading in mS^{-1} or μS cm^{-1}.

Notes

1. The soil sample is collected as described in experiment no. 1
2. The distilled water is boiled before use. This minimises the CO_2 content. The electrical conductance must be less than 0.01 mS m^{-1}.

(c) Result

The conductivity of the soil sample is........... mS m^{-1} or μS cm^{-1}.

(d) Inference

On the basis of the determation of conductance of soil it can be known that which soil is good for crop growth (non-Saline Soil, conductance < 0.40 mS m^{-1} at 25°), slightly saline (conductance 0.40 – 0.80 mS m^{-1} at 25°C, such soils are considered stafe for most crops, moderately saline (conductance 0.81–1.20 mS m^{-1} at 25°C; such soils are useful only for salt-tolerant corps. For other crops, reclamation is necessary) and saline (conductance > 1.21 mS m^{-1}; such soil need reclamation for the growth of crops).

Experiment 5

Determination of the Presence or Absence of Different Water Soluble Cations and Anions in Soil Sample.

The water soluble cations which may be present in soil are ammonium (NH_4^+), ferrous (Fe^{2+}), ferric (Fe^{3+}), manganese (Mn^{++}), calcium (Ca^{2+}), magnesium (Mg^{2+}).

The anions which may be present in soil are chloride (Cl^-), sulphate (SO_4^{2-}), nitrate (NO_3^-), phosphate (PO_3^{3-}).

Following are given procedures for testing the presence of various cations and anions in the soil.

Ammonium (NH_4^+)

It can be directly tested in the soil by heating with sodium hydroxide solution. Smell of ammonia indicates the presence it NH_4^+. The presence of ammonia can be confirmed by passing the evolved gas into nessler's reagent, when brown colour or precipitate is obtained

$$NH_4X + NaOH \xrightarrow{\Delta} NH_3\uparrow + H_2O + NaX$$

$$\underset{\text{Nesslers reagenl}}{K_2HgI_4} \longrightarrow 2KI + HgI_2$$

$$HgI_2 + NH_3 \longrightarrow Hg\begin{matrix}\diagup NH_2\\ \diagdown I\end{matrix} + HI$$

$$2Hg\begin{matrix}\diagup NH_2\\ \diagdown I\end{matrix} + 2H_2O \longrightarrow \underset{\text{Brown precipitate}}{\left[O\begin{matrix}\diagup Hg \diagdown\\ \diagdown Hg \diagup\end{matrix}NH_2\right]} I.H_2O + NH_4^+ + I^-$$

The Nessler's reagent required for the above test is prepared by dissolving 5g Potassium iodide in minimum amount of water and adding saturated mercuric chloride solution until a faint permanent precipitate is obtained. To the mixture is added 40 mL of 50 per cent aqueous potassium hydroxide solution. The mixture is allowed to stand and then diluted is 100 mL with distilled water. The resulting mixture is allowed to stand overnight and the clear solution is decanted. The clear decanted solution is the Nessler's reagent.

Iron (Fe^{2+}, Fe^{3+})

The soil sample (about 20g) is stirred with 5 mL dilute hydrochloric acid and filtered. To the filtrate is added ammonium chloride (1 to 2g) and sodium hydroxide solution (3–4 mL) and hydrogen per oxide (3–4 mL, 6 per cent). The solution is boiled and then filtered. The residue (X) (Containing hydroxides of iron and manganese) is tested for the presence of iron.

To part of the residue, is added dilute hydrochloride (2mL) and divided in two parts

$$Fe(OH)_3 + 3HCl \longrightarrow FeCl_3 + 3H_2O$$

To one part of the solution is added few drops of KCNS solution. Formation of blood red colour confirms the presence of Fe^{3+}.

$$FeCl_3 + 3KCNS \longrightarrow \underset{\text{Blood red colour}}{Fe\,(CNS)_3} + 3KCl$$

To the other part of the solution is added few drops of $K_4Fe\,(CN)_6$ Solution. Formation of a blue precipitate or colour confirms Fe^{3+}

$$4FeCl_3 + 3K_4\,[Fe\,(CN)_6] \longrightarrow \underset{\text{Blue colour}}{Fe_4\,[Fe\,(CN)_6]_3} + 12KCl$$

Manganese (Mn^{2+})

The residue (X) obtained in the testing of iron is washed with water and dissolved conc. HNO_3. The solution is boiled and treated with 3–4 drops of hydrogen peroxide (6 percent). The solution is boiled again and solid sodium bismuthate (0.1 to 0.2g) or solid PbO_2 (0.1–0.2g) added. The mixture is allowed to settle. Formation of violet or purple colour confirms the presence of Mn^{2+}.

$$Mn(OH)_2 + 2HNO_3 \longrightarrow Mn(NO_3)_2 + 2H_2O$$

$$2Mn(NO_3)_2 + 5\,PbO_2 + 6\,HNO_3 \longrightarrow \underset{\text{(Violet)}}{2HMnO_4} + 2H_2O + 5Pb\,(NO_3)_2$$

Calcium (Ca^{2+})

To about 25g of dry soil sample is added 50 mL dilute ethyl alcohol (40%). The mixture is shaken (4–5 min) and allowed to stand (10 min.) and filtered. The residue is washed with dilute alcohol (40%) and then with absolute alcohol. The remaining residue is suspended in 50mL. ammonium acetate solution stirred and allowed to stand over night. The mixture is filtered and the filtrate concentrated to a small volume (10–15mL).

To the concentrated solution is added solid NH_4Cl (2–3g) and NH_4OH solution (4–5 mL). To the resulting solution, is added saturated solution of ammonium carbonate. The resuling mixture is filtered [The filtrate (A) is used for testing the presence of magnesium]. To the precipitate is added excess of ammonium sulphate solution. The solution warmed and treated with ammonium oxalate solution. Formation of white precipitate confirms the presence of Ca^{2+}.

$$CaSO_4 + (NH_4)_2\ SO_4 \longrightarrow (NH_4)_2\ [Ca(SO_4)_2]$$

$$\underset{}{(NH_4)_2\ [Ca\ (SO_4)_2] + (COONH_4)_2} \longrightarrow \underset{\text{white}}{(COO)_2\ Ca\downarrow} + 2(NH_4)_2\ SO_4$$

The white precipitate obtained above is dissolved is conc. HCl and flame test performed. Brick red colour in the flame confirms the presence of Ca^{2+}.

Magnesium (Mg^{27}).

The filtrate (A) obtained above in the testing for calcium is concentrated to dryness. The residue is heated strongly in a fume cupboard to decompose ammonium salts. The cooled residue is dissolved in dil. HCl (1–2mL), heated and filtered (if necessary). To the clear solution is added few drops of Na_2HPO_4 and few drops of NH_4OH till the solution smells of ammonia. Scratch the inner side of the test tube with a glass rod. Formation of white precipitate shows the presence of Mg^{2+}.

$$MgCl_2 + Na_2HPO_4 + NH_4OH \longrightarrow \underset{\text{(white ppt.)}}{Mg\ (NH_4)\ PO_4\downarrow} + 2NaCl + H_2O$$

Chloride (Cl^-)

A small amount (about 1–2g) of the dry soil sample is heated with conc. H_2SO_4 (1–2mL). Evolution of a colourless, pungent smelling HCl gas indicates the presence of Cl^- (a glass rod dipped in ammonia solution brought near the mouth of the test tube gives dense white fumes of NH_4Cl with HCl gas)

$$NaCl + H_2SO_4 \longrightarrow NaHSO_4 + 4HCl$$

$$HCl + NH_4OH \longrightarrow NH_4Cl + H_2O$$

Alternatively, the soil is suspended in dilute HNO_3 and then filtered. To the filtrate is added 0.5 mL $AgNO_3$ solution. Formation of a white precipitate of AgCl (soluble in dilute ammonia solution) confirms the presence of Cl^-.

$$AgNO_3 + NaCl \longrightarrow \underset{\text{white ppt.}}{AgCl\downarrow} + NaNO_3$$

$$AgCl + 2NH_4OH \longrightarrow \underset{\text{Soluble Complex}}{[Ag(NH_3)_2]\ Cl} + 2H_2O$$

Sulphate (SO_4^{2-})

The soil sample (about 5g) is heated with water (10mL) and filtered. To the clear filtrate is added $BaCl_2$ solution (1 – 2mL). Formation of white precipitate indicates the presence it SO_4^{2-}

$$\underset{x\,=\,\text{Ca, Mg, NH}_4\text{. etc.}}{X_2SO_4 + BaCl_2} \longrightarrow \underset{\text{white ppt.}}{BaSO_4\downarrow} + XCl$$

Nitrate (NO_3^-)

The soil sample (about 5g) is heated with water (5–10mL) and filtered. To the filtrate is added solid $FeSO_4$ (0.5g) followed by addition of conc. H_2SO_4 (1mL) through the sides of the test tube. Formation of a brown ring at the junction of the two liquids confirms the presence of NO_3^-.

$$NaNO_3 + H_2SO_4 \longrightarrow NaHSO_4 + HNO_3$$

$$2HNO_3 + 6FeSO_4 + 3H_2SO_4 \longrightarrow 3Fe_2(SO_4)_3 + 2NO + 4H_2O$$

$$FeSO_4 + NO \longrightarrow \underset{\text{Brown ring or colour}}{FeSO_4.NO}$$

Phosphate (PO_4^{3-})

The soil sample (about 5g) is treated with dilute HNO_3 (1mL) and filtered. To the filtrate is added ammonium molybdate solution (1–2mL). A yellow precipitate is formed on heating

$$Na_3PO_4 + 24HNO_3 + 12\,(NH_4)_2\,MoO_4 \xrightarrow{\Delta} 3NaNO_2 + 21NH_4\,NO_3 + 12H_2O + \underset{\text{Yellow ppt.}}{(NH_4)_3\,PO_4.12\,MoO_3}$$

Experiment 6

Determination of Hardness of Water

We know that hardness of water is of two types:

(*i*) **Temporary hardness (or carbonate hardness):** It is caused by the presence of carbonates and bicarbonates, and can be removed by boiling water. Such waters give leather with soap.

(*ii*) **Permanent hardness (or non-carbonates hardness):** It is caused by the presence of calcium and magnesium salts (chlorides, sulphates, nitrates) and cannot be removed by boiling. Such waters do not give leather with soap.

In fresh water, the hardness is caused mainly by the presence of Mg^{2+} and Ca^{2+}. However, cations of iron, aluminium, manganese, strontium and zinc also cause hardness, but to a very slight extent. The total hardness is defined as the sum of calcium and magnesium ions present. It is expressed as ppm of $CaCO_3$. The degree of hardness is classified in terms of equivalent $CaCO_3$ concentration as moderately hard or very hard having equivalent $CaCO_3$ concentration of 75–150 ppm or >150 ppm, respectively. Water having equivalent of $CaCO_3$ concentration 0–75 ppm is soft water.

Basis of determination of hardness of water

Ethylene diamine tetra acetic acid (EDTA) reacts with both Ca^{++} and Mg^{++} in alkaline conditions (pH 10.0 ± 0.1) to form a soluble chelated complex. Addition of few drops of an indicator, Eriochrome Black T (a dye) to the aqueous solution containing Ca^{++} and Mg^{++} gives wine red colour. On titration against standard EDTA solution, the Ca^{++} and Mg^{++} ions form complex when all the Ca^{++} and Mg^{2+} ions have formed complex (at the end point), the solution turns blue from wine red. The pH for this titration must be maintained at 10.0 ± 0.1.

In case, the pH is higher (about 12.0), Mg^{2+} ions get precipitated and only Ca^{2+} ions remain in solution. At this pH, addition of Murexide indicator and titration against standard EDTA solution, the pink colour due to Ca^{2+} ion changes to purple colour at the end point. In case other metal ions are present, these do interfere, but this interference can be overcome by addition of inhibitor (a solution of 2.25g hydroxylamine hydrochloride in 50 mL of 95 per cent ethyl alcohol)

Procedure

(*a*) Total hardness

To 50 mL of water sample (in a conical flask) is added 1–2 mL of a buffer solution[1] and 1mL of inhibitor[2]. Using a pinch of Eriochrome Black T indicator[3], the solution is titrated against standard EDTA (0.01M) Solution[4] till the wine red colour changes to blue.

Repeat the titration, till two concordant readings are obtained (Let the volume of EDTA used be V_1)

Run a reagent blank and note the volume of EDTA used (Let the volume be V_2)

$\therefore$ the volume of EDTA required for the sample $V_3 = (V_1 - V_2)$

Total hardness as $CaCO_3$ (mg/L) is determined by using the formula $\frac{V_3 \times D \times 1000}{\text{Volume of sample (ml)}}$ where V_3 is the volume of EDTA required by the sample

D = mg. of $CaCO_3$ equivalent to 1.0 mL of EDTA titrant.

Notes

1. The buffer solution is prepared by dissolving 32g of NH_4Cl in 285 mL of ammonia solution (Sp.gr. 0.88) and adding 2.358g of disodium salt of EDTA (molar mass = 372.31g mol^{-1}) and 1.560g of $MgSO_4.7H_2O$ or 1.788g of Mg $Cl_2.6H_2O$ in 100 mL of distilled water and making the volume to 500 mL with distilled water.
2. The inhibitor is a solution of 2.25g hydroxylamine hydrochloride in 50 mL of ethyl alcohol.
3. Eriochrome Black T indicator solution is prepared by mixing 0.05g indicator and 100g NaCl.
4. EDTA solution (0.01m) is prepared by dissolving 1.862g EDTA disodium salt in distilled water and diluting to 500 mL.

(*b*) Calcium hardness

To 50 mL of water sample (in a conical flask) is added 1mL 2NNaOH solution (obtained by dissolving 40g NaOH in distilled water and making the volume to 500 mL) to raise the pH to 12.0 and a pinch of murexide indicator (a grounded mixture of 0.2g of murexide (ammonium purpurate) with 100g solid sodium chloride). The solution is titrated against standard EDTA solution (0.01m) and volume of EDTA required is noted. Repeat the titration, till two concordant readings are obtained (Let the volume of EDTA used be V_1').

Run the reagent blank and note the volume of EDTA used (Let to volume be V_2')

∴ The volume of EDTA required for the sample $V_3' = (V_1' - V_2')$

$$\text{Calcium hardness} = \frac{V_3' \times D \times 1000}{\text{Volume of sample (mL)}}$$

where V_3' is the volume of EDTA used by the sample (mL)

D = mg. of $CaCO_3$ equivalent to 1.0 mL of EDTA

titrant (0.01M) = 0.4mg of Ca = 1mg of $CaCO_3$

Magnesium hardness = Total hardness (as $CaCO_3$) – Ca hardness (as $CaCO_3$)

Experiment 7

Estimation of Calcium in Soil

Calcium is a major constituent of soil. It is also present in water and is responsible for hardness of water. The calcium finds its way into soil and water bodies by washings (run offs) from the deposits of limestone, dolomite and gypsum.

Calcium can be estimated by EDTA titration procedure. It can also be calculated from calcium hardness (sec experiment no.6). The multiplying factor is 0.4 when hardness is expressed as $CaCO_3$.

Procedure

(*a*) Preparation of Soil Extract

To 50g of air-dry soil (taken in a conical flask) add 100 mL of 40 per cent ethyl alcohol. Shake the suspension (5–10 min) and allow to settle (15 min). The suspension is filtered using whatman filter paper no. 50. The residue on the filter paper in washed first with 40 per cent ethyl alcohol and then with absolute alcohol.

The residual soil (in the filter paper) is transferred to a beaker and stirred with 100 mL ammonium acetate solution (1N) and allowed to stand overnight. The supernatant liquid is filtered through whatman filter paper no.50. The filtrate known as soil extract is used for the estimation of calcium.

(*b*) Estimation of Calcium

To 50 mL of soil extract (taken in a conical flask) is added 1 mL sodium hydroxide solution (8%) and a pinch of murexide indicator (prepared as in experiment no.6).

The solution is titrated against 0.1m EDTA solution (prepared as in experiment no.6) until the pink colour turns purple at the end point.

$$\text{Calcium hardness (mg/L)} = \frac{V \times D \times 100}{\text{Volume of sample (ml)}}$$

where V = mg of $CaCO_3$ EDTA used by soil extract (mL)

D = mg of $CaCO_3$ equivalent to 1.0 mL. of EDTA titrant.

Experiment 8

Estimation of Magnesium in Soil

Magnesium is an important constituent of soil. It is also responsible for hardness of water like calcium. The magnesium also finds its way into land and water bodies by the washings (run offs) from ores containing magnesium. It is an essential constituent of chlorophyll.

Magnesium can be estimated by EDTA titration procedure. It can also be estimated as the difference between total hardness and calcium hardness. The appropriate factor to convert the difference to Mg is 0.444 if both total and calcium hardness are expressed as $CaCO_3$.

Procedure

(*a*) Preparation of soil cxtract

It is prepared as in case of calcium estimation (experiment no. 7).

(*b*) Estimation of Magnesium

As a first step total hardness in soil extract is determind as in the case of water (experiment no.7). From these values, magnesium amount is calculated as given below.

$$\text{Mg (mg/g)} = (\text{Th} - \text{Ch}) \times 0.244$$

(1g equiv. of EDTA = 24g Mg)

where Th is the total hardness (mg/litre, as $CaCO_3$)

and Ch is the calcium hardness (mg/litre as $CaCO_3$)

Magnesium (mg/g) =

$$\frac{(V_3 - V_3') \times 400.4 \times \text{total volume of soil extract}}{\text{Volume of soil extract used for titration} \times 1.645 \times W \times 10{,}000}$$

where V_3 is the volume of EDTA used for determination of Ca and Mg (mL)

V_3' is the volume of EDTA used for determination of calcium (mL)

W is the weight of dry soil taken for preparation is of soil extract (g)

Experiment 9

Estimation of Oxidisable Organic Matter in Soil

The oxidisable organic matter in soil can be determined by titration of soil extract against standard ferrous ammonium sulphate solution using diphenylamine as indicator

Procedure

To 0.5g of air dry soil sample (taken is a 500 mL conical flask) is added 10 mL of 1N potassium dichromate solution followed by gradual addition of 20mL conc. sulphuric acid. The flask is allowed to stand for 1 hr. During this time, the organic matter is oxidised.

To the solution is added 200 mL distilled water, 10mL phosphoric acid, 0.2g sodium fluoride (dry) and 1mL of diphenyl amine indicator.

The solution is titrated against 0.5N ferrous ammonium sulphate. At the end point, the light green colour changes to brilliant green.

A blank experiment is run with distilled water. Calculate the percentage of oxidisable organic matter as follows:

$$\text{Organic matter (mg/g)} = \frac{6.791}{W}\left[1-\frac{V_1}{V_2}\right]\times 10$$

$$\text{Percentage of organic matter} = \frac{6.791}{W}\left[1-\frac{V_1}{V_2}\right]$$

where W is the mass of the soil (g)

V_1 is the volume of ferrous ammonium sulphate solution used for titration of soil extract.

V_2 is the volume of ferrous ammonium sulphate solution in blank experiment.

Experiment 10

Determination of the Biological Oxygen Demand (B.O.D.) of a Given Water Sample.

Theory

There exists a close relation between organic matter and dissolved oxygen (DO) in water, which is expressed in terms of BOD. BOD is a measure of the amount of oxygen required by bacteria and other microorganisms.

The BOD is defined as the standardized measurement of the amount of oxygen required by the microorganisms for the biological oxidation or decomposition of organic matter under aerobic conditions to CO_2 and H_2O at 20°C over a period of 5 days as 70 to 80% of organic matter is oxidized during this period. The result is called the five day BOD and is expressed in milligrams of oxygen per litre of water (mg/litre). The oxidation of organic matter is represented as.

$$\underset{\text{Organic matter}}{CH_2O} + \underset{\text{(DO)}}{O_2} \xrightarrow[\text{Aerobic Oxidation + Bacterial Cell}]{\text{Microorganism}} CO_2 + H_2O$$

This method consists of measuring the amount of oxygen consumed by a sample of known volume. This method involves the collection of sample in an air tight bottle and incubating it at specified temperature for five days. DO is the basis of BOD test. DO is measured initially and after incubation, and BOD is computed from the difference between initial and final DO.

Reagents

1. Phosphate buffer.
2. Magnesium Sulphate solution (22.5g/l).
3. Calcium Chloride solution (27.5g/l)
4. Ferric Chloride solution (0.25 g/l)
5. 0.025N Sodium Thiosulphate solution.

Procedure

(*i*) A stock quantity of dilution water is prepared by taking a desired volume of distilled water in a suitable bottle and pH is adjusted to 7–7.2 with the addition of phosphate buffer, 1mL of $MgSO_4$, 1mL of $CaCl_2$ and 1mL of $FeCl_3$.

(*ii*) The above water is thoroughly aerated by bubbling air through a diffusion tube until the dissolved oxygen level is 7ppm.

(*iii*) Take two samples of known quantity of sewage water in 300mL bottle and adjust the pH to 7 with 1N H_2SO_4 or 1N NaOH.

(*iv*) Make a measured dilution of the samples with dilution water prepared as above (*i*) if BOD is greater than the DO level at 7ppm. At this stage, it is necessary to add micro-organism to the sample under investigation in order to obtain a valid BOD. This is known as seeding.

(*v*) Seed the samples with a little (1–2mL) diluted domestic waste water. Dilute the sample further with water to the brim of BOD bottle.

(*vi*) Keep these BOD bottles incubated in the dark at 20°C for the 5 days.

(*vii*) The dissolved oxygen in the first bottle is determined immediately without incubation and is taken is D_1, which gives the initial dissolved oxygen concentration of sample.

(*viii*) After five days, take second sample bottle and measure the dissolved oxygen of sample taken as D_2 which gives the dissolved oxygen present after incubation.

(*ix*) During the course of incubation organic matter get decomposed. The unused dissolved oxygen at different times of incubation is determined by titrating the acidified 1N H_2SO_4 sample solution with sodium thiosulphate (0.25N) until it is colourless. Sodium thiosulphate in acid converts to dithionic acid which reacts with dissolved oxygen to form sodium dithionate. Each mL of thiosulphate used in titration is equivalent to 1mg/l of DO.

(*x*) Similarly, take two samples of dilution water called control samples and saturate them by bubbling with oxygen and then seed with microorganisms.

(*xi*) The first control sample is acidified and titrated with hypo solution to get the dissolved oxygen before incubation (B_1).

(*xii*) The second control sample after five days is acidified and titrated with hypo solution to get the amount of dissolved oxygen after incubation (B_2).

The reaction involved are given below:

$$Na_2S_2O_3 + H_2SO_4 \longrightarrow H_2S_2O_3 + Na_2SO_4$$

$$2H_2S_2O_3 + O_2 \longrightarrow H_2S_2O_6 + H_2O$$

$$H_2S_2O_6 + Na_2SO_4 \longrightarrow Na_2S_2O_6 + H_2SO_4$$

$$Na_2S_2O_3 + O_2 \longrightarrow Na_2S_2O_6 + H_2O$$

Calculations

From the decrease in concentration of DO, the BOD can be calculated by using the following expression:

$$\text{BOD in mg/l} = \frac{\text{Number of mg of } O_2 \text{ required}}{\text{Number of litres of sample}}$$

$$\text{BOD} = \frac{(D_1 - D_2) - (B_1 - B_2)f}{P}$$

Where D_1 is DO of diluted sample before the incubation of sewage sample.

D_2 is DO of diluted sample after the incubation of the sewage sample

B_1 and B_2 are DO levels of diluted water of control samples containing seeds before and after incubation.

P = decimal fraction of seeds (microorganisms) added to the sample after investigation.

f = ratio of seeds in the control sample.

The difference between DO in diluted water and unused oxygen of the solution after five days gives the BOD value. The BOD of 100mg/l means the biodegradation of organic matter in a liter of sample, consumes 100mg of oxygen.

Result:- The BOD of given water sample is..........mg/l.

Experiment 11

Determination of Chemical Oxygen Demand (COD) of a Given Water Sample.

Theory:-

COD is the parameter to determine the total amount of oxygen matter in streams, industrial waters and sewage. It is an index to the water oxygen demand of substances in water. The chemical oxygen required for determining the amount of organic matter in water is determined by $K_2Cr_2O_7$. $K_2Cr_2O_7$ is a strong chemical oxidizing agent which is capable of oxidizing all organic matter present in water by chemical oxidation.

COD is defined as the amount of oxygen required by organic matter in a sample of water for its oxidation by a strong oxidizing agent like $K_2Cr_2O_7$ in the presence of $HgSO_4$ and Ag_2SO_4 catalysts. It is expressed as the ppm of oxygen taken from a $K_2Cr_2O_7$ solution in two hours.

The reactions involved are given below:

$$3CH_2O + 16H^+ + 2Cr_2O_7^{2-} \longrightarrow 4Cr^{3+} + 3CO_2 + 11H_2O$$

$$Cr_2O_7^{2-} + 14H^+ + 6e^- \longrightarrow 2Cr^{3+} + 7H_2O$$

$$O_2 + 4H^+ + 4e^- \longrightarrow 2H_2O$$

1 mole of $Cr_2O_7^{2-}$ consumed is equal to (6/4) 1.5 moles of oxygen liberated and used for oxidation. COD is a better parameter than BOD for evaluating the amount of organic matter present in water. For the given organic pollutant, the COD will always be greater than that of its BOD values. This is because in COD, both the biologically oxidizable and the biologically inert matter are oxidized. The main advantage of COD is that its determination takes just 3 hours compared to five days for BOD determination.

The determination of the COD of a system depends upon the determination of those compounds that are capable of undergoing chemical oxidation. The decomposition of these materials reduce the available oxygen within a system. The amount of organic matter oxidised is found out by titration with ferrous ammonium sulphate and the COD is calculated.

Procedure

(*i*) Take 20mL of the sample and add 30mL of dilute H_2SO_4.

(*ii*) Add 10mL of standard potassium dichromate solution (0.25N) and 2–3 boiling chips and reflux for 2 hours.

(*iii*) The solution is cooled and diluted with distilled water to make 150mL.

(*iv*) Add a few drops of ferroin indicator (O-phenanthroline + $FeSO_4$) solution and unreacted $K_2Cr_2O_7$ solution is titrated to a reddish brown end point with the standardized ferrous ammonium sulphate (0.1N).

(*v*) A blank titration with 20mL of distilled water in place of the sample is to be titrated in the same manner.

Calculation

The amount of oxidizable organic matter present in the sample in terms of COD value can be calculated by using the following equation which directly gives the COD value in mg/litre.

$$\text{COD} = \frac{(a - b) \times c \times 8000}{\text{Volume of sample taken (ml)}}$$

Where

a = volume of $Fe(NH_4)_2\ (SO_4)_2$ used for blank.

b = volume of $Fe\ (NH_4)_2\ (SO_4)_2$ used for sample.

c = volume of $Fe(NH_4)_2\ (SO_4)_2$ determined by the standardization with $K_2Cr_2O_7$.

Result:- The COD of given water sample is ....mg/litre.

Experiment 12

Determination of the Amount of Sodium Carbonate and Sodium Bicarbonate Present in the Given Solution Using Hydrochloric Acid (N/20)

Theory

The reaction between Na_2CO_3 and HCl takes place in the following two steps. Na_2CO_3 is first converted to $NaHCO_3$.

$$Na_2CO_3 + HCl \longrightarrow NaHCO_3 + NaCl \qquad ...(i)$$

and then the sodium bicarbonate reacts with HCl.

$$NaHCO_3 + HCl \longrightarrow NaCl + CO_2 + H_2O \qquad ...(ii)$$

Sodium bicarbonate present in the mixture also reacts (equation (*ii*)) with HCl to form NaCl and CO_2.

The strength of Na_2CO_3 in the given solution can be determined by titrating with standardized HCl in presence of phenolphthalein as the indicator. The solution (pink) becomes colourless when Na_2CO_3 is converted to $NaHCO_3$ (equation (*i*)). Thus, the volume of HCl used in this titration when multiplied by two will be equivalent to the total volume of HCl required to neutralize the total amount of sodium carbonate in the solution.

The second set of titration is carried out using methyl orange as the indicator; methyl orange changes colour from yellow to red when whole of Na_2CO_3 and $NaHCO_3$ have been neurtalized.

Procedure

(*i*) **Standardisation of HCl:** Weight accurately ~ 0.26 g of anhydrous sodium carbonate and dissolve to make 100 mL solution in a standard flask.

Pipette out 20 mL of the standard sodium carbonate in a conical flask and add 1–2 drops of methyl orange. The solution acquires a yellow colour. Titrate against hydrochloric acid slowly till the yellow solution becomes orange–red. Repeat the titration to obtain three concordant readings. (let the volume of HCl be V_1).

(*ii*) **Titration of $Na_2CO_3 + NaHCO_3$ using Std. HCl:** Now pipette out 20 mL of the given solution of Na_2CO_3 and $NaHCO_3$ in a conical flask, add a drop of phenolphthalein (the solution acquires a pink colour) and titrate against HCl till a colourless solution is obtained (let the volume of HCl be V_2). Repeat the titration to obtain three concordant readings.

Next pipette out 20 mL of the given solution of sodium carbonate + sodium bicarbonate, add 1–2 drops of methyl orange and titrate with HCl till a red colour is obtained (let the volume of HCl be V_3). Repeat to obtain three concordant readings.

Calculations

Mass of Na_2CO_3 in 100 mL solution = y g

Normality (N) of standard sodium carbonate $= \frac{y \times 10}{53} = N$

Normality (N_1) or HCl is $N_1V_1 = NV$

$$\underset{\text{(HCl)}}{N_1V_1} = \underset{\text{(Standard } Na_2CO_3)}{NV}$$

$$N_1 = \frac{NV}{V_1} = N_1$$

Volume of HCl used for neutralization of 20 mL of given solution of ($Na_2CO_3 + NaHCO_3$) when phenolphthalein is used as an indicator = V_2

∴ Volume of HCl required to neutralize the total amount of Na_2CO_3 present in the given solution of ($Na_2CO_3 + NaHCO_3$) = $2V_2$

∴ Normality N_2 of Na_2CO_3 in the given solution ($Na_2CO_3 + NaHCO_3$)

$$N_2 \times 20 = 2V_2 \times N_1$$

(Na_2CO_3 in given solution) (Standardized HCl)

∴ Strength of $Na_2CO_3 = N_2$

$$= \frac{2V_1N_1}{20} = N_2$$

$$\text{Amount of } Na_2CO_3 = \frac{2 \times 2V_2N}{20} \times 53\text{g/liter}$$

Volume of HCl required to neutralize total amount of Na_2CO_3 + $NaHCO_3$ present in the given solution = V_3 (*i.e.*, when methyl orange is used as indicator)

So, the volume of HCl required to neutralize $NaHCO_3$ only in the solution = $V_3 - 2V_2$

Normality (N_3) of $NaHCO_3$

$$N_3 \times 20 = N_1(V_3 - 2V_2)$$

$$N_3 = \frac{N_1(V_3 - 2V_2)}{20};$$

Where N_3 is the Normality of $NaHCO_3$ in the given solution

Amount of $NaHCO_3$ in the given solution = $N_3 \times 84$ g/litre

$$= \frac{N_1(V_3 - 2V_2) \times 84}{20} \text{ g/liter}$$

❒❒❒

POST SCRIPT
OZONE PACT DIDN'T SLOW WARMING

Replacements for CFCs help Antarctic hole but trap heat just the same

Associated Press

WASHINGTON—Cool your home, warm the planet.

When more than two dozen countries undertook in 1989 to fix the ozone hole over Antarctica, they began replacing chlorofluorocarbons in refrigerators, air conditioners and hairspray.

But they had little idea that using other gases, *containing chloring or fluorine instead, also would contribute greatly to global warming*

CFCs destroy ozone, the atmospheric layer that helps protect against the sun's most harmful rays, and trap the Earth's heat, contributing to a rise in average surface temperatures, scientists largely agree.

In theory, the ban should have lessened both problems. But the countries that first signed the Montreal Protocol 17 years ago failed to recognize that CFC users would seek out the cheapest alternatives.

The chemicals that replaced CFCs are better for the ozone layer, but do little to counter global warming. These chemicals, too, act as a reflective layer in the atmosphere that traps heat like a greenhouse, research has shown.

That effect is at odds with the intent of a second treaty, drawn up in Kyoto, Japan, in 1997 by the same countries behind the Montreal pact. In fact, the volume of greenhouse gases created as a result of the Montreal agreement's phaseout of CFCs is two to three times the amount of global—warming carbon dioxide that Kyoto is supposed to eliminate.

This unintended consequence now haunts the nations.

Switzerland first tried in 1990 to sound an alarm that the solution of plugging the ozone hole might contribute to another environmental problem. The reason?

"Nothing, or almost," said Blaise Horisberger, the Swiss representative to the UN—backed Montreal treaty. Mr. Horisberger, a biologist with the Swiss Agency for the Environment, Forests and landscape, Kept trying. Finally, formal, secret talks on the subject were held in Montreal.

The Montreal Protocol, which now has 189 member nations, is considered one of the most effective environmental treaties. Almost $2.1 billion has been spent through an affiliated fund to request countries to stop making and using CFCs and other ozone—damaging chemicals in refrigerators, air conditioners, foams and other products.

The majority of scientists blame CFCs for poking a huge, seasonal hole in the stratospheric ozone layer, about 7 to 14 miles over Antarctica. In 2005, the ozone hole peaked at about 10 million square miles, or the size of North America.

CFCs also are thinning the ozone layer over the Arctic and, to a lesser extent, over the Arctic and, to a lesser extent, globally, say scientists. As the protective layer thins, they say, more ultraviolet radiation gets through.

Some of the replacement chemicals whose use has grown because of the Montreal treaty hydrochlorofluorocarbons or **HCFCs,** and their byproducts, hydro fluorocarbons, or **CFCs** because they contain hydrogen.

But, *like CFCs, they are considered potent greenhouse gases that harm the climate—up to 10,000 times worse than carbon dioxide emissions.*

The *Kyoto treaty's* goal is to reduce carbon dioxide emissions from fossil—fuel—burning power plants, motor vehicles and other sources by *about 1 billion tons by 2012*.

Use of HCFCs and HFCs is projected to add the equivalent of *2 billion to 3 billion tons of carbon dioxide emissions* to the atmosphere by 2015, U.N. climate experts said in a recent report. The CFCs they replace also would have added that much.

The U.N. report says the atmosphere could be spared if countries used ammonia., hydrocarbons, carbon dioxide or other ozone friendly chemicals, rather than HCFCs and HFCs, in foams and refrigerants. Such alternatives are more common in Europe, but industry is split over how to replace CFCs and HCFCs.

Industry representatives cite safety and energy—efficiency problems with using ammonia and hydrocarbons, which mainly involves propane gas.

Manufacturers also say they couldn't meet U.S energy—efficiency requirements using those chemicals.

Note: The above write up is on the basis of a report published in the Daily Morning News, Dallas, Texus, USA dated Aug. 21,2006.

❐❐❐

Index

A

B

C

D

E

F

G

H

I

J

K

L

M

N

❐❐❐